Peter Westermann

SIMD Processing for Software Defined Radio

Peter Westermann

SIMD Processing for Software Defined Radio

Exploration of the Scalability of SIMD Processing for Software Defined Radio

Südwestdeutscher Verlag für Hochschulschriften

Imprint
Any brand names and product names mentioned in this book are subject to trademark, brand or patent protection and are trademarks or registered trademarks of their respective holders. The use of brand names, product names, common names, trade names, product descriptions etc. even without a particular marking in this work is in no way to be construed to mean that such names may be regarded as unrestricted in respect of trademark and brand protection legislation and could thus be used by anyone.

Publisher:
Südwestdeutscher Verlag für Hochschulschriften
is a trademark of
Dodo Books Indian Ocean Ltd., member of the OmniScriptum S.R.L Publishing group
str. A.Russo 15, of. 61, Chisinau-2068, Republic of Moldova Europe
Printed at: see last page
ISBN: 978-3-8381-2675-3

Zugl. / Approved by: Dortmund, TU, Diss, 2011

Copyright © Peter Westermann
Copyright © 2011 Dodo Books Indian Ocean Ltd., member of the OmniScriptum S.R.L Publishing group

Acknowledgements

This book covers my doctoral thesis, which has been written during my stay with the Circuits and Systems Lab (CaS Lab) at the TU Dortmund University.

First of all I would like to thank my advisor Prof. em. Dr.-Ing. Hartmut Schröder for the many fruitful discussions and the opportunity to be engaged in challenging research topics. The success of this thesis would not have been possible without him. I'm also very grateful for experiencing the unique and pleasant working atmosphere at the CaS Lab.

I also would like to thank Prof. Dr.-Ing. Holger Blume from the Institute of Microelectronic Systems at the Leibniz University Hannover for co-supervising my thesis and his interest in general.

This thesis evolved from research projects on SIMD processing with Nokia and later Nokia Siemens Networks, all under the guidance of Prof. Dr.-Ing. Ludwig Schwoerer from the Communication and Electronics Institute at the Bochum University of Applied Sciences. Thanks a lot for the support and contributions to this thesis — and for making me focus on SIMD processing and SDR systems instead of image processing.

My thanks also goes to John Thompson and Xiang Wu at the University of Edinburgh for the helpful discussions on sphere decoding and inviting me to Edinburgh.

Furthermore, I would like to say thanks to all my colleagues, former colleagues and students at the TU Dortmund University and my project partners with Nokia and Nokia Siemens Networks.

Above all, I would like to thank my family for love and support.

Abstract

The idea of software defined radio (SDR) describes a signal processing system for wireless communications that allows performing major parts of the physical layer processing in software. SDR systems are more flexible and have lower development costs than traditional systems based on application-specific integrated circuits (ASICs). Yet, SDR requires programmable processor architectures that can meet the throughput *and* energy efficiency requirements of current third generation (3G) and future fourth generation (4G) wireless standards for mobile devices.
Single instruction, multiple data (SIMD) processors operate on long data vectors in parallel data lanes and can achieve a good ratio of computing power to energy consumption. Hence, SIMD processors could be the basis of future SDR systems. Yet, SIMD processors only achieve a high efficiency if all parallel data lanes can be utilized.
This work investigates the scalability of SIMD processing for algorithms required in 4G wireless systems; i. e. the scaling of performance and energy consumption with increasing SIMD vector lengths is explored. The basis of the exploration is a scalable SIMD processor architecture, which also supports long instruction word (LIW) execution and can be configured with four different permutation networks for vector element permutations.
Radix-2 and mixed-radix fast Fourier transform (FFT) algorithms, sphere decoding for multiple input, multiple output (MIMO) systems, and the decoding of quasi-cyclic low-density parity check (LDPC) codes have been examined, as these are key algorithms for 4G wireless systems. The results show that the performance of all algorithms scales with the SIMD vector length, yet there are different constraints on the ratios between algorithm and architecture parameters. The radix-2 FFT algorithm allows close to linear speedups if the FFT size is at least twice the SIMD vector length, the mixed-radix FFT algorithm requires the FFT size to be a multiple of the squared SIMD width. The performance of the implemented sphere decoding algorithm scales linearly with the SIMD vector length. The scalability of LDPC decoding is determined by the expansion factor of the quasi-cyclic code. Wider SIMD processors offer better performance and also require less energy than processors with a shorter vector length for all considered algorithms. The results for different permutations networks show that a simple permutation network is sufficient for most applications.

Contents

Abstract	II
Table of contents	IV
List of figures	IX
List of tables	XII
Notation	XV
1 Introduction	**2**
2 Overview of software defined radio principles and architectures	**6**
2.1 Software defined radio	6
2.1.1 Reconfigurable SDR architectures	7
2.1.2 SIMD-based architectures for SDR	8
2.2 Basic principles of SIMD processing	9
2.2.1 SIMD vector processing	9
2.2.2 Advantages and disadvantages of SIMD processing	11
2.3 Wide SIMD processor architectures and research on the scalability of SIMD processing	11
2.3.1 The Embedded Vector Processor	12
2.3.2 The Sandblaster SB3500 architecture	14
2.3.3 The Signal-processing On-Demand Architecture	15
2.3.4 The Ardbeg architecture based on SODA	17
2.3.5 Processor architectures based on SIM_dD processing	18
2.3.6 Research on the scalability of SIMD processing for SDR	20
2.4 Key algorithms for future 4G SDR systems	23
2.4.1 MIMO-OFDM system model	24
3 Scalable SIMD processor architecture	**26**
3.1 Development of the SIMD processor architecture based on algorithm requirements	26
3.1.1 Word lengths and data types	27
3.1.2 Instruction set	28
3.1.3 Instruction level parallelism	31
3.1.4 Register files	34
3.1.5 Permutation networks	38

	3.1.6	Overview of the SIMD processor model	43
3.2	SIMD processor modeling in LISA		45
	3.2.1	Processor Designer toolkit overview	46
	3.2.2	Processor modeling in LISA	46
	3.2.3	Extensions for modeling SIMD processors	47
	3.2.4	Drawbacks of LISA as a modeling language for SIMD processors	49
3.3	Vertical-horizontal vector processing as an alternative for LIW		51
	3.3.1	Vertical-horizontal vector processing for SDR	52
	3.3.2	SDR algorithm performance	53
3.4	SIMD architecture analysis methodology		54
	3.4.1	Processor model synthesis	54
	3.4.2	Extraction of area, power, energy and performance figures	56
	3.4.3	Limitations of the proposed methodology	56

4 Radix-2 and mixed-radix FFTs for OFDM-A and SC-FDMA 59

4.1	OFDM-A and SC-FDMA		59
4.2	Matrix representation of the FFT		61
	4.2.1	Basic DFT decomposition for two factors	61
	4.2.2	Formula manipulation rules for the DFT in matrix form	62
	4.2.3	Vectorizable formulas	63
4.3	Related work on SIMD FFT algorithms		64
4.4	Derivation of SIMD radix-2 and mixed-radix FFT algorithms		65
	4.4.1	Short radix-2 FFT algorithm	65
	4.4.2	Mixed-radix FFT algorithm	69
	4.4.3	Permutations for the vectorized FFT algorithms	73
4.5	Radix-2 and mixed-radix FFT implementations based on LTE		79
	4.5.1	Grouping of FFT stages	80
	4.5.2	Implementation of DFT stages	83
	4.5.3	Implementation of permutation stages for different permutation networks	84
	4.5.4	Short mixed-radix FFT implementation	86
4.6	Performance analysis		88
	4.6.1	Overview of throughput results	88
	4.6.2	Speedup results	88
	4.6.3	Resource utilization and performance of FFT loops	91
	4.6.4	Comparison to other SDR FFT implementations	93
4.7	Conclusion		95

5 Sphere decoding for MIMO detection 96

5.1	MIMO system model		96
	5.1.1	Maximum likelihood detection	98
	5.1.2	Sphere decoding	98

		5.1.3 Soft-decision MIMO detection	100

- 5.2 Breadth-first search MIMO decoders ... 101
  - 5.2.1 The K-best sphere decoder ... 101
  - 5.2.2 Selective spanning with fast-enumeration ... 102
- 5.3 The fixed-complexity sphere decoder ... 104
  - 5.3.1 FSD tree search ... 104
  - 5.3.2 FSD ordering of the channel matrix ... 105
  - 5.3.3 Soft-decision MIMO detection based on the FSD ... 106
- 5.4 SIMD implementation of the FSD for MIMO-OFDM ... 108
  - 5.4.1 Channel ordering ... 110
  - 5.4.2 QR-decomposition by Givens rotations ... 114
  - 5.4.3 Hard-decision FSD tree search ... 116
  - 5.4.4 Soft-decision FSD tree search extension by bit-flipping ... 117
  - 5.4.5 LLR calculation for soft-decision MIMO decoding ... 118
- 5.5 Performance analysis ... 119
  - 5.5.1 Overview of FSD results ... 119
  - 5.5.2 Analysis of the achievable throughput ... 121
  - 5.5.3 Comparison to SDR and hardware-based sphere decoders ... 124
  - 5.5.4 Improving the FSD performance ... 126
- 5.6 Conclusion ... 127

6 Decoding of quasi-cyclic low density parity check codes   128

- 6.1 Fundamentals ... 128
  - 6.1.1 Definition of LDPC codes ... 128
  - 6.1.2 Representation by Tanner graphs ... 129
  - 6.1.3 Quasi-cyclic LDPC codes ... 129
- 6.2 Decoding of LDPC codes ... 131
  - 6.2.1 Decoding schedules ... 131
  - 6.2.2 Iterative decoding algorithms ... 133
- 6.3 SIMD implementation of LDPC decoding for WiMAX ... 135
  - 6.3.1 Algorithm for min-sum decoding ... 136
  - 6.3.2 Implementation for the parallel processing of one sub-matrix ... 138
  - 6.3.3 Implementation for the parallel processing of multiple sub-matrices ... 139
- 6.4 Performance analysis ... 142
  - 6.4.1 Throughput and speedup results ... 142
  - 6.4.2 LIW resource utilization ... 144
  - 6.4.3 Comparison to other architectures ... 145
  - 6.4.4 Improving the LDPC decoding performance ... 146
- 6.5 Conclusion ... 148

7 Evaluation of the SIMD architecture efficiency — 149
- 7.1 Area and power consumption results 149
  - 7.1.1 Average power consumption 150
  - 7.1.2 Area 151
  - 7.1.3 Power consumption and area estimates for memories 152
- 7.2 Energy efficiency analysis 153
  - 7.2.1 Normalized energy consumption 154
  - 7.2.2 Energy-delay product analysis 158
- 7.3 Possible approaches for improving the scalability 160
  - 7.3.1 Vector alignment with indirect SIM_dD processing 160
  - 7.3.2 Support for operations on vector segments 162
- 7.4 Software development for LIW SIMD processors 163

8 Conclusion — 166

Bibliography — 172

List of Figures

2.1	SIMD-based SoC for SDR	8
2.2	Examples for SIMD vector operations	9
2.3	Example SIMD processor architecture with four lanes	10
2.4	Block diagram of the EVP architecture	12
2.5	Block diagram of SODA	15
2.6	Topologies of perfect shuffle exchange and inverse perfect shuffle exchange networks	15
2.7	Simplified block diagram of a SIMD computation unit in SODA-II	16
2.8	Conceptual view of the vector pipeline in SODA-II	17
2.9	Block diagram of the SIMD and scalar data paths of an Ardbeg PE	18
2.10	Indirect SIM_dD processing	19
2.11	Block diagram of an AnySP processing element	20
2.12	Block diagram of transmitter and receiver in a MIMO-OFDM system	24
3.1	Definitions of arithmetic data types	29
3.2	Visualization of ILP architectures based on [Smo02]	32
3.3	Fixed-length and variable-length LIW encoding	33
3.4	Encoding of 24-bit slots	35
3.5	Read/write connections between the general-purpose SIMD register file and the SIMD units	36
3.6	8-point DIF FFT	37
3.7	4×4 crossbar network based on transmission gates	39
3.8	Three MIN topologies based on cube network (a), Omega network (b) and butterfly network (c)	41
3.9	Example demonstrating the limited permutation support of MINs	42
3.10	Block diagram of the scalable SIMD processor architecture	44
3.11	Pipeline of the scalable SIMD processor architecture	45
3.12	Example of an operation hierarchy with a four-stage pipeline	48
3.13	Example for vertical-horizontal vector processing with two vector units	51
3.14	Cyclic mapping of FUs on register banks	52
3.15	Block diagram showing the synthesis and analysis methodology	54
3.16	Dynamic power consumption	57
4.1	Block diagram of an SC-FDMA transmitter	60
4.2	Signal flow graphs corresponding to Kronecker products with DFT matrices	64

List of Figures

4.3 Signal-flow graph for a vectorized 8-point FFT for a SIMD width of four complex-valued elements. 67
4.4 Stride permutation defined by $\mathbf{P}_8^3$ for a vector length of four elements 68
4.5 Permutations in between radix-2 FFT stages for the mixed-radix FFT with $m = 3$ and $V = 4$ 69
4.6 Block diagram of a 48-point mixed-radix FFT for a vector length of four 72
4.7 Stride permutation (a) based on $\mathbf{P}_4^2$ for $V = 4$ and realization on a single-vector permutation network (b) 76
4.8 Masked butterfly permutations with different block sizes for a vector length of eight .. 77
4.9 8-point SIMD FFT algorithm with butterfly permutations 78
4.10 Assembly code for realizing the equivalent of a masked 64-bit butterfly permutation on a pair of vectors by permutations on a single-vector network 85
4.11 Merging of two FFTs for virtually reducing the vector length 87
4.12 Speedup for radix-2 and mixed-radix FFTs on different SIMD processors measured versus a 128-bit SIMD processor with a single-vector butterfly network 92

5.1 Channel model for a MIMO system with n_T transmit and n_R receive antennas 96
5.2 Sphere search in a 4×4 MIMO system for QPSK 100
5.3 K-best tree search in a 4×4 MIMO system for QPSK 102
5.4 4×4 MIMO SSFE search trees for two different spanning vectors 103
5.5 Fast enumeration of the two closest nodes based on distance to computed ξ_i for 16-QAM modulation 104
5.6 FSD tree search for QPSK and a 4×4 MIMO system 105
5.7 LFSD tree search for QPSK and a 4×4 MIMO system with $N_S = 8$ 107
5.8 Example search tree for FSD with bit-flipping for generating additional paths (4×4 MIMO with QPSK modulation) 109
5.9 Parallel processing of the FSD algorithm by parallel processing of OFDM sub-carriers for a SIMD width of four elements 110
5.10 Example for the reduction of matrix $\mathbf{A}$ during the channel ordering 112
5.11 Swapping of data vectors v0 and v1 based on vector mask m1 using two parallel masked move operations 113
5.12 Merging of data vectors for the parallel computation of reciprocal square roots. 116
5.13 Symbol detection by threshold comparisons in the scaled constellation diagram for 16-QAM 118
5.14 Best-case 4×4 MIMO FSD throughput for different SIMD widths for hard-decision (a) and soft-decision (b) decoding 122
5.15 Assembly code fragment for the calculation of the squared Euclidean distance 126
5.16 Assembly code fragment for the thresholding operation during SE stages for 16-QAM modulation 126

6.1 Tanner graph of a (7,4) Hamming code 130

List of Figures

6.2   LDPC decoding with (a) flooding schedule and (b) check node serial schedule 132
6.3   Check node serial iterative decoding algorithm for $M \times N$ LDPC codes 134
6.4   Message calculation for message from bit node n to check node m using the min-sum algorithm 137
6.5   Check node update based on incoming message for min-sum algorithm 137
6.6   Cyclic shift by two for $z = 12$ and $V = 4$ 138
6.7   Parallel processing of sub-matrices in a row of $\mathbf{H}_b$ for $z = 4$, $V = 8$ 140
6.8   Merging of code words for parallel processing 141
6.9   Cyclic right shift by two for $z = 12$ and $V = 8$ for block-interleaved parallel processing of code words. 141
6.10 Speedup of LDPC decoding compared to a 128-bit SIMD processor with a single-vector inverse butterfly network 143
6.11 Calculation of $E_{m,n}^i$ during the message-passing 147
6.12 Update of minimum and minimum position during the check node update 148

7.1   Normalized power consumption 151
7.2   Normalized area 152
7.3   Speedup of algorithms compared to 128-bit SIMD processor with single-vector inverse butterfly network 158
7.4   Normalized energy consumption of the implemented algorithms 159
7.5   Normalized energy-delay product and area for radix-2 and mixed-radix FFTs 160
7.6   Normalized energy-delay product and area for soft-decision and hard-decision FSD . 161
7.7   Normalized energy-delay product and area for the decoding of quasi-cyclic LDPC codes 162
7.8   Cyclic shift operation on an inverse butterfly network. The complete cyclic shift is computed from cyclic shifts on smaller vector segments. 163

8.1   Processor utilization for the example MIMO-OFDM transmission scenario 169
8.2   Total energy consumption for the example MIMO-OFDM transmission scenario 170
8.3   Normalized energy-delay product and area for the described scenario 171

List of Tables

2.1	Overview of commercial and academical SIMD processors for SDR	13
2.2	Analysis of data parallelism [WLS+08b]	22
3.1	Supported basic arithmetic data types	28
3.2	Vector processing units and supported operation types	30
3.3	Scalar processing units and supported operation types	31
3.4	Latencies of scalar and vector instructions measured in clock cycles	31
3.5	Measured ILP on the EVP for inner loops of baseband algorithms	34
3.6	Evaluation of the grouping of radix-2 butterfly stages	38
3.7	Register files in the scalable SIMD processor architecture	38
3.8	Summary of properties of the four implemented networks	43
3.9	Design space for the exploration of the scalability of SIMD processing	43
3.10	Source lines of code measured for LISA model before and after M4 macro expansion	50
3.11	Model-based register file comparison of monolithic and partitioned register files	53
3.12	Performance comparison of LIW and vertical-horizontal vector processing architectures	53
3.13	SDR algorithms used for power optimization during synthesis	55
3.14	Standard cell library operating conditions	55
4.1	IDFT sizes at the transmitter side in LTE [Tec10]	61
4.2	Constraints for radix-2 and mixed-radix FFT sizes for different vector lengths	79
4.3	Implemented radix-2 and mixed-radix FFTs	80
4.4	Short radix-2 FFTs that fit into the vector register file	80
4.5	Decomposition of long radix-2 and mixed-radix FFTs into groups of FFT stages in loops	81
4.6	Overview of additional permutation stages for short radix-2 FFTs	86
4.7	Peak throughput in FFTs per second without overhead for initialization	89
4.8	Throughput in FFTs per second with overhead for initialization (long CP mode)	90
4.9	Overview of the LIW performance of FFT loops	93
4.10	Comparison of SDR implementations of radix-2 and mixed-radix FFTs	94
5.1	Number of searched paths and node distribution vectors for the LFSD and a 4×4 MIMO system with 16-QAM modulation	107
5.2	Number of candidates for the soft-decision FSD with bit-flipping at each tree level for 4×4 MIMO	108
5.3	Complexity comparison of scalar and vector channel matrix reordering for a 4×4 matrix	114
5.4	Polynomial coefficients for reciprocal square root approximation	115

List of Tables

- 5.5 Overview of FSD implementation results ... 120
- 5.6 4×4 FSD throughput on 128-bit SIMD processor architectures ... 121
- 5.7 Throughput requirements for different channel bandwidths for 4×4 MIMO with 16-QAM modulation ... 123
- 5.8 Maximum number of resource blocks that can be decoded in real-time for 4×4 MIMO with 16-QAM modulation ... 124
- 5.9 Performance of sphere decoding algorithms with fixed-complexity in ASICs and on FPGA 124
- 5.10 Performance of sphere decoding algorithms with fixed-complexity on SDR processors ... 125

- 6.1 Notation for LDPC decoding ... 133
- 6.2 Implemented WiMAX codes [IEE09b] ... 135
- 6.3 Block matrices for WiMAX [IEE09b] ... 135
- 6.4 Notation for min-sum decoding ... 136
- 6.5 Decomposition of check node/bit node update operations and cyclic shifts into loops ... 139
- 6.6 Required memory for permutation patterns and masks for cyclic shifts on segments of vectors on a 1024-bit SIMD processor with an inverse butterfly network ... 141
- 6.7 Throughput of LDPC decoding with 10 decoding iterations ... 142
- 6.8 LIW resource utilization of bit node and check node update loops ... 145
- 6.9 Overview of LDPC decoder implementations for WiMAX ... 146

- 7.1 Area and power consumption results for the scalable SIMD processor architecture ... 150
- 7.2 Permutation network area ... 153
- 7.3 Area of processor core and memories ... 154
- 7.4 Power consumption of processor core and memories ... 155
- 7.5 Total energy consumption for the implemented algorithms on 128-bit and 256-bit SIMD processors ... 156
- 7.6 Total energy consumption for the implemented algorithms on 512-bit and 1024-bit SIMD processors ... 157
- 7.7 Comparison of compiler generated code on the EVP and hand-optimized assembly code on the proposed SIMD architecture (256 bit SIMD width, crossbar network). FFT throughput is measured in FFTs per second, while FSD throughput is measured in OFDM sub-carriers per second ... 165

Notation

Abbreviations

3G	Third Generation
4G	Fourth Generation
ADL	Architecture Description Language
AGU	Address Generation Unit
ALU	Arithmetic Logic Unit
APP	A Posteriori Probability
ASIC	Application-Specific Integrated Circuit
ASIP	Application-Specific Instruction-set Processor
ASP	Application-Specific Processor
BER	Bit Error Rate
BP	Belief Propagation
BPSK	Binary Phase Shift Keying
BU	Branch Unit
CDMA	Code Division Multiple Access
CGU	Code Generation Unit
CP	Cyclic Prefix
DFT	Discrete Fourier Transform
DIF	Decimation In Frequency (FFT)
DL	DownLink
DMA	Direct Memory Access
DSE	Design Space Exploration
DSP	Digital Signal Processor or Digital Signal Processing
DVB-H	Digital Video Broadcasting - Handhelds
DVB-T	Digital Video Broadcasting Terrestrial
EPIC	Explicitly Parallel Instruction Computing
EVP	Embedded Vector Processor
EVP-C	EVP C language
FE	Full Expansion (stage of the FSD algorithm)
FEC	Forward Error Correction
FFT	Fast Fourier Transform
FFU	Flexible Functional Unit
FIR	Finite Impulse Response

FPGA	Field Programmable Gate Array
FSD	Fixed-Complexity Sphere Decoder
FU	Functional Unit
GMAC/s	Giga MAC operations per second
GMul./s	Giga Multiplications per second
GOPS	Giga Operations Per Second
GPP	General-Purpose Processor
HSDPA	High Speed Downlink Packet Access, extension of UMTS
IDFT	Inverse Discrete Fourier Transform
IFFT	Inverse Fast Fourier Transform
i. i. d.	Independent Identically-Distributed
ILP	Instruction Level Parallelism
IMT-Advanced	International Mobile Telecommunications-Advanced
IP	Intellectual Property
ISA	Instruction Set Architecture
ISI	InterSymbol Interference
ISS	Instruction Set Simulator
ITRS	International Technology Roadmap for Semiconductors
IVU	Intra Vector Unit
LDPC	Low-Density Parity Check
LFSD	List Fixed-complexity Sphere Decoder
LISA	Language for Instruction Set Architecture
LIW	Long Instruction Word
LLR	Log-Likelihood Ratio
LMMSE	Linear Minimum Mean Squared Error
LSU	Load/Store Unit
LTE	Long Term Evolution
LTE-Advanced	Long Term Evolution-Advanced
LUT	LookUp Table
MAC	Multiply-Accumulate or Multipy-Accumulate Unit
MALU	Mask ALU
MAP	Maximum A posteriori Probability
MIMO	Multiple Input, Multiple Output
MIN	Multistage Interconnect Network
ML	Maximum Likelihood
MMSE	Minimum Mean Squared Error
MOPS	Million Operations Per Second
MSB	Most Significant Bit
NMOS	N-channel Metal–Oxide–Semiconductor (field-effect transistor)
nop	No-Operation
OFDM	Orthogonal Frequency Division Multiplex

Notation

OFDM-A	Orthogonal Frequency Division Multiple Access
PALU	Predicate ALU
PAPR	Peak-to-Average Power Ratio
PCU	Program Control Unit
PDA	Personal Digital Assistant
PE	Processing Element
PED	Partial Euclidean Distance
PiCoGA	Pipelined Configurable Gate Array
PMOS	P-channel Metal–Oxide–Semiconductor (field-effect transistor)
QAM	Quadrature Amplitude Modulation
QPSK	Quadrature Phase Shift Keying
QRD-M	QR-Decomposition and M-algorithm
RCA	Reconfigurable Communication Architecture
RF	Register File
RISC	Reduced Instruction Set Computer
RTL	Register Transfer Level
SC-FDMA	Single Carrier Frequency Division Multiple Access, also denoted as DFT-spread OFDMA
SD	Sphere Decoder
SDR	Software Defined Radio
SE	Single Expansion (stage of the FSD algorithm)
SFSD	Soft-decision FSD algorithm with bit flipping
SIMD	Single Instruction (stream), Multiple Data (streams)
SIM$_d$D	Single Instruction (stream), Multiple disjoint (or distributed) Data (streams)
SIMO	Single Input, Multiple Output
SIMT	Single Instruction stream, Multiple Task
SISO	Single Input, Single Output
SLOC	Source Lines Of Code
SNR	Signal-to-Noise Ratio
SoC	System on a Chip
SODA	Signal-processing On-Demand Architecture
SRAM	Static Random-Access Memory
SSFE	Selective Spanning with Fast Enumeration
SSN	SIMD Shuffle Network
STBC	Space Time Block Coding/Code
SVD	Singular Value Decomposition
SXU	Scalar Exchange Unit
TD-SCDMA	Time Division Synchronous Code Division Multiple Access
TTA	Transport-Triggered Architecture
UL	UpLink
UMTS	Universal Mobile Telecommunications System

VALU	Vector ALU
V-BLAST	Vertical Bell Laboratories Layered Space-Time
VLIW	Very Long Instruction Word
VLSU	Vector LSU
VMAC	Vector MAC Unit
VMU	Vector Move Unit
VPU	Vector Permutation Unit
VSHU	Vector Shuffle Unit
W-CDMA	Wideband Code Division Multiple Access
WiMAX	Worldwide Interoperability for Microwave Access
WLAN	Wireless Local Area Network
ZOL	Zero-Overhead Loop, also denoted as hardware loop

Symbols and mathematical notation

$\det \lvert \cdot \rvert$	Matrix determinant operator
$\operatorname{sign}(\cdot)$	Sign operator
$\max(\cdot)$	Maximum operator
$\min(\cdot)$	Minimum operator
$\operatorname{P}(\cdot)$	Probability
$\operatorname{Re}\{\cdot\}$	Real part of a complex value
$\operatorname{Im}\{\cdot\}$	Imaginary part of a complex value
$\lfloor \cdot \rfloor$	Operator for rounding down
$\mathcal{O}(\cdot)$	Landau notation for asymptotic complexity
$Q.f$	Fixed-point data type with one sign bit, zero integer bits and f fractional bits; example: $Q.15$
$Qm.f$	Fixed-point data format with one sign bit, m integer bits and f fractional bits; example: $Q8.31$
$\mathbf{A}$	Matrix
$\mathbf{v}$	Vector
i	Imaginary unit $i = \sqrt{-1}$
$A_{i,j}$	Element of matrix $\mathbf{A}$ in row i and column j. Alternatively, the matrix element operator $[\mathbf{A}](i,j)$ is also used.

Notation

$[\mathbf{A}](i,j)$	Matrix element operator
$\mathbf{I}_N$	N-point identity matrix
$\mathbf{A}^T$	Transpose of matrix $\mathbf{A}$
$\mathbf{A}^H$	Conjugate transpose of complex-valued matrix $\mathbf{A}$
a^*	Conjugate complex of scalar value a
$\det(\mathbf{A})$	Determinant of matrix $\mathbf{A}$
$\operatorname{adj}(\mathbf{A})$	Adjugate matrix of matrix $\mathbf{A}$
$\|\cdot\|_2$	Euclidean (matrix) norm
$\|\cdot\|$	Absolute value
$\otimes$	Kronecker product operator: $\mathbf{A} \otimes \mathbf{B} = (a_{ij} \cdot \mathbf{B})$
$\oplus$	Bitwise exclusive or operator
SNR	Signal-to-noise ratio (SNR)
N_0	Noise power
$\mathbf{W}_N$	N-point DFT matrix
N_{DFT}	DFT size/length
$\operatorname{GF}(q)$	Galois field with q elements
$\lceil \cdot \rceil$	Operator for rounding up to the next greater integer number

Chapter 1
Introduction

The idea of software defined radio (SDR) has gained significant interest in the last decade. SDR describes a wireless communication device, where *"some or all of the physical layer functions are software defined"* [For07]. The term radio characterizes any kind of wireless communication device that transmits or receives radio frequency signals. Examples of radio devices range from cell phones, laptop computers, and PDAs to door openers and navigation systems. An SDR system comprises programmable signal processing devices that allow implementing some or all of the physical layer operations in software. These devices may be based on field programmable gate arrays (FPGAs), digital signal processors (DSPs), application-specific processors (ASPs) or multiple of these devices in a system on chip (SoC).

SDR systems offer some advantages in comparison to solutions based on application-specific integrated circuits (ASICs) [SVPG+10]: a common programmable architecture can be used for multiple different wireless protocols and products. This allows reducing the time to market for new products and the development costs, as software and hardware can be reused, leading to increasing chip volumes. Programmability also reduces maintenance costs, as software updates and bug fixes can be applied either over-the-air or by other means of remote programming. The possibility of software updates also enables future extensions to existing radio systems. A common platform for multiple protocols enables service providers to use one platform for multiple markets and enables multi-mode operation for the customers, i. e., depending on the available network technologies, different protocols can be used (e. g. UMTS, LTE, WiMAX, or WLAN) without need for additional hardware.

Yet, the benefits of SDR come with steep requirements: the radio device has to operate in real-time with low power consumption. This is especially true for battery-powered mobile devices, where an efficiency of approximately 100 million operations per second (MOPS) per milliwatt is necessary for third generation (3G) wireless protocols [Lin08] with even higher demands for future fourth generation (4G) systems [WLS+08b, Rep08]. The power budget for baseband signal processing is approximately 500 mW [Lin08]. Hence, there is a need for efficient processor architectures for SDR – with parallel processing as the only means to achieve the required efficiency.

Programmable architectures for SDR can be classified into two groups — reconfigurable architectures and architectures based on single instruction, multiple data (SIMD) processing [Ram07]. Coarse-grain and fine-grain reconfigurable architectures (e. g. [MVV+03, LCB+06]) offer less flexibility than SIMD architectures and the programming model for SIMD architectures is simpler than the programming model of reconfigurable architectures. Hence, SIMD processor architectures are especially interesting for future 4G systems.

SIMD processor architectures contain multiple parallel data lanes that all execute the same instruction on different elements of long data vectors. The combination of parallel processing and a low overhead for the decoding of instructions leads to a good efficiency. SIMD processor architectures can be categorized into wide SIMD architectures with 16 or more parallel data lanes [vHM+05, LLW+06, WLS+08a, MG08, WSM+09] and short SIMD architectures with few parallel data lanes [GI06, Kno05, GRS07, Nil07]. The number of parallel lanes is denoted as the SIMD width or vector length.

Future SDR devices based on SIMD processing will be designed as SoCs with multiple SIMD processor cores for tasks that can be efficiently parallelized and require programmability and few ASICs as accelerators for tasks that cannot be efficiently mapped on a SIMD processor and/or do not require programmability. Such an SDR SoC could comprise many small or few wide SIMD processor cores. In principle, wide SIMD processors can achieve better energy efficiency than processors that only support short SIMD operations [WLS+08a]. Yet, a high performance and a good energy efficiency can only be achieved for algorithms that can utilize all parallel data lanes efficiently. The optimal SIMD width depends on architecture parameters as well as algorithm parameters. While SIMD processing is a technique that has been thoroughly investigated, few research results have been published that discuss the issue of selecting an appropriate SIMD width for SDR [WLS+08a, WLS+08b].

Objective of this book

This book focuses on the issue of the scalability of SIMD processing for SDR algorithms: How do performance and energy consumption of a SIMD processor architecture scale with an increasing SIMD width for the key algorithms of future 4G systems?

In order to be able to evaluate different SIMD widths, a scalable SIMD processor architecture has been developed using the LISA language [ZPM96, PHZM99, Pee02] and synthesized in a 90 nm technology for SIMD widths ranging from 8 to 64 parallel 16-bit lanes. The processor architecture combines SIMD processing with long instruction word (LIW) execution — different processing units can execute operations in parallel to each other — to reduce the overhead for data alignment and memory access operations. Furthermore, four different permutation networks for vector element permutations have been modeled.

Three key algorithms for future 4G systems have been mapped on the scalable SIMD processor architecture and their performance has been explored. The considered algorithms are radix-2 and mixed-radix FFTs, which are required for orthogonal frequency division multiplex (OFDM) systems, a sphere decoding algorithm for multiple input, multiple output (MIMO) systems, and the decoding of quasi-cyclic low-density parity check (LDPC) codes.

The major contributions of this paper are as follows:

- A scalable SIMD processor architecture has been developed in LISA. The processor architecture enables a systematic exploration of a design space that comprises four SIMD widths and four different permutation networks. As the processor architecture also supports LIW execution, the mitigation of overhead for memory access and vector alignment by LIW has also been investigated.

- The mapping of three state-of-the-art algorithms for 4G MIMO-OFDM systems on the scalable SIMD processor architecture has been investigated. The scalability of the algorithms has been studied and constraints for the scalability on SIMD processors have been identified. The radix-2 FFT algorithm requires the FFT size to be at least twice the SIMD width, the mixed-radix FFT algorithm requires an FFT size that is a multiple of the squared SIMD width. The sphere decoded implementation processes multiple OFDM sub-carriers in parallel, which enables efficient vectorization for SIMD widths up to the OFDM symbol size. The decoding algorithm for quasi-cyclic LDPC codes can efficiently utilize SIMD widths up to the value of the expansion factor of the code. Wider SIMD widths can also be utilized using a modified algorithm for the parallel processing of multiple sub-matrices.

- An algorithm for mixed-radix FFTs has been developed that enables the vectorization of FFT sizes that are a multiple of the squared SIMD width. An efficient mapping of permutation stages on a fixed number of simple butterfly permutations is developed.

- The performance of the implemented algorithms and the energy efficiency on the proposed scalable SIMD architecture have been investigated. The results show that linear or close to linear speedups can be achieved for all algorithms, as long as the constraints for scalability are satisfied. The energy efficiency increases with the SIMD width. The exploration of different permutation network topologies shows that wider permutation networks may lead to performance gains for FFT processing, but wider permutation networks also require more register file ports, which leads to significantly higher register file power consumption. The achieved throughputs of FFTs and sphere decoding are sufficient for data rates beyond 100 Mbps; LDPC decoding requires further modifications on the architecture to meet the requirements of future 4G systems.

Synopsis

Chapter 2 introduces SDR and SIMD-based processor architectures for SDR. The advantages and disadvantages of SDR and SIMD processing are discussed and relevant wide SIMD architectures and related work on the scalability of SIMD processing are reviewed. Furthermore, the key algorithms for future 4G systems are identified.

In chapter 3, the development of the proposed scalable SIMD processor architecture is discussed in detail. The used modeling language LISA is introduced and the methodology for the evaluation of the SIMD processor architecture is discussed. Moreover, an alternative for LIW processing based on vertical-horizontal vector processing is briefly sketched.

Chapter 4 discusses radix-2 and mixed-radix FFTs that are required for orthogonal frequency division multiple access (OFDM-A) and single carrier frequency division multiple access (SC-FDMA), which is an extension of OFDM-A used for the uplink channel in the long term evolution (LTE) standard [Tec09b]. A novel decomposition for the mixed-radix FFT, which is optimized for SIMD processing, is derived from the matrix representation of the FFT and constraints for the efficient SIMD processing of radix-2 and mixed-radix FFTs are developed. Implementations of FFTs on the proposed scalable SIMD processor architecture are discussed and performance results, which show approximately linear speedups, are presented.

In chapter 5, different algorithms for MIMO detection based on the sphere decoder (SD) are explained. The fixed-complexity sphere decoder (FSD) is discussed in detail as it has been implemented on the scalable SIMD processor architecture. The FSD has good properties for parallel processing and achieves a bit error rate (BER) close to the maximum likelihood (ML) solution. The algorithm implementation is explained and the throughput performance is analyzed and compared to other SDR and hardware MIMO detection implementations.

The following chapter (chapter 6) focuses on the decoding of LDPC codes. The properties of LDPC codes and decoding algorithms based on message-passing are discussed. It is shown that quasi-cyclic LDPC codes — as e. g. optionally supported by WiMAX [IEE09b] — are the most promising codes from an implementation point of view, because of their regular structure. A parallel implementation of quasi-cyclic LDPC decoding for WiMAX is discussed and limitations of the proposed scalable SIMD processor architecture are demonstrated. The chapter concludes with a performance analysis and a comparison to other SDR and hardware implementations.

In chapter 7, the scalable SIMD processor architecture is analyzed based on the results obtained for radix-2 and mixed-radix FFTs, MIMO detection, and LDPC decoding. The chapter contains an analysis of area, power consumption, and energy efficiency of processors with different SIMD widths and permutation networks. Furthermore, modifications on the processor architecture that could improve the scalability of mixed-radix FFTs and LDPC decoding are explained. The chapter concludes with a brief discussion of the complexity of mapping algorithms on SIMD processors and the possible support of this task by vectorizing compilers.

Finally, chapter 8 contains a summary of results, conclusions for the scalability of SIMD processing, and an outlook of possible future research directions.

Chapter 2

Overview of software defined radio principles and architectures

This chapter contains an overview of SIMD processing and the concept of software defined radio (SDR). Section 2.1 discusses advantages and disadvantages of SDR and gives an overview of SDR architecture classes. Section 2.2 describes SIMD vector processing and discusses its advantages and disadvantages. Wide SIMD processor architectures and related research on the scalability of SIMD processing are described in section 2.3. The chapter concludes with an overview of the key algorithms of future 4G systems/standards in section 2.4.

2.1 Software defined radio

The SDR principle describes a wireless communication device that implements the physical layer processing exclusively or mostly on programmable processing architectures, for example digital signal processors (DSPs), application-specific processors (ASPs), or reconfigurable architectures. Classic architectures for wireless communication are based on application-specific integrated circuits (ASICs), assisted by one or few DSPs [Ram07], and do not offer any flexibility. The programmability of SDR solutions offers some advantages versus the traditional hardware solutions [SVPG+10]:

- SDR solutions enable performing software development and hardware design and verification in parallel. Furthermore, new protocols can be quickly implemented in software and mapped onto an existing SDR platform, without a hardware redesign. Therefore, development time and costs are significantly lower than for classical ASIC solutions. Development costs are further reduced by the potentially higher chip volumes, as one SDR platform can be used for different applications.

- SDR solutions enable dynamically executing different wireless communication protocols on the same processor architecture (multi-mode operation). Commonly used functions, such as filters, encoders/decoders, and transforms (FFT), can be adjusted at runtime. Multi-mode capability allows service providers to use one platform for different markets and offer more functionality to the end users. Furthermore, hardware costs are reduced if resources can be shared effectively between wireless protocols.

- The programmability of SDR systems also reduces maintenance costs, as new functionality (e. g. new standard releases) or bug fixes can be applied either over-the-air or by other means of remote programming.

The keynote of these arguments is that SDR increases the flexibility and reduces the costs of wireless communication devices. Yet, the increased flexibility has its price, as SDR systems will always consume more power than an optimized ASIC-centered solution [Ram07]. The chip area of a programmable SDR architecture for one wireless protocol is also greater than the chip area using ASICs, but this can be canceled out by multi-mode support, i. e. multiple wireless protocols are supported on the same SDR architecture, while an ASIC-centered approach would require adding further ASICs, which increases the chip area.

One further argument sometimes used against programmable SDR solutions is the notion that many of the algorithms in the physical layer processing of wireless communication protocols could be efficiently realized on dedicated hardware accelerators with limited reconfigurability, e. g. allowing parameter adjustments for filters. However, this approach leads to an increased development time and increasing costs for designing accelerators that can support the requirements of multiple wireless protocols. Furthermore, an accelerator will only support one algorithm. Some critical physical layer tasks, such as MIMO symbol detection, can be performed by many different (possibly in-house) algorithms, which achieve different trade-offs between algorithm complexity (e. g. runtime, required memory) and algorithm performance (e. g. BER). An SDR architecture that offers sufficient computing power can perform any of these algorithms, allowing companies to implement their own preferred solution.

Future 4G wireless protocols aim at data rates between 100 Mbps and 1 Gbps [Rep08, Tec09a]. Therefore, SDR solutions need to achieve high throughputs, while still complying with the power restrictions of wireless communication devices. The power budget for baseband processing in mobile devices is approximately 500 mW (see [Lin08, Neu04]), with a power efficiency of approximately 100 million operations per second (MOPS) per milliwatt required for 3G and even steeper requirements for 4G. Therefore, energy efficiency is of essential importance for SDR systems. High energy efficiency can be achieved by parallel processing and by employing application-specific instructions or processing units. Programmable architectures for SDR can be categorized into two philosophies — reconfigurable architectures and architectures based on SIMD processors [Ram07].

2.1.1 Reconfigurable SDR architectures

The design-flow for reconfigurable SDR architectures is as follows: First, algorithms or parts of algorithms that are used by multiple or all targeted wireless protocols are identified. Next, reconfigurable data paths, which provide the necessary processing for this common functionality, are designed. The flexibility of reconfigurable SDR architectures depends on the granularity of the decomposition into common data paths. An example for a fine-grain reconfigurable data path is a small DSP core, which implements elementary functions, such as addition or multiplication. Coarse-grain reconfigurable data paths might implement complete algorithms, for example, a data path might be realized as an ASP for FFT processing.

Examples for fine-grain reconfigurable SDR architectures are University of Bologna's XiRisc processor architecture [LCB+06] and picoChip's picoArray [Pul08, BDT08]. The XiRisc processor is a VLIW RISC processor, with two data paths with arithmetic and DSP-like functional units and a third data path based on a pipelined configurable gate array (PiCoGA) [LTC03]. Application-specific functions can be mapped onto the PiCoGA. The picoArray architecture consists of many independent RISC processors organized in a two-dimensional array. For example, the picoArray PC102 comprises 308 processors. Each processor executes its own instruction stream and processes its own data. Processors are connected by a high-speed time-division multiplexed bus system.

Intel's reconfigurable communication architecture (RCA) [CTC+04] and IMEC's ADRES architecture [MVV+03, SVPG+10] are examples for coarse-grain reconfigurable SDR architectures. Intel's RCA consists of a mesh of heterogeneous, reconfigurable processing elements (PEs), connected by routers. The PEs are optimized for different parts of the baseband processing, e.g. filtering or Turbo coding. The ADRES architecture combines a VLIW processor with a coarse-grained reconfigurable array. The coarse-grained array consists of reconfigurable functional units (FUs) with local memories. Neighboring FUs can exchange data without any register file accesses in between. The VLIW core and the FUs communicate through a global register file and shared memory. A typical ADRES instance consists of a 4×4 array [SVPG+10] with 128-bit SIMD FUs (12 reconfigurable units and 4 units in the VLIW part of the architecture).

2.1.2 SIMD-based architectures for SDR

SIMD-based architectures for SDR utilize SIMD processor cores to achieve high energy efficiency. The basics of SIMD processing are discussed in section 2.2. A system on a chip (SoC) for SDR based on SIMD processing consists of many SIMD processor cores and few ASICs [Ram07]. The ASICs accelerate algorithms that do not require programmability and/or cannot be efficiently mapped on the SIMD processors. Figure 2.1 shows a block diagram of a SIMD-based SoC for SDR.

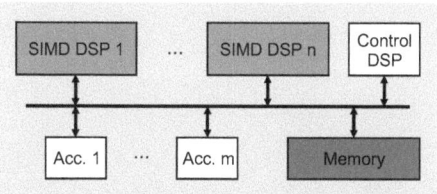

Figure 2.1: SIMD-based SoC for SDR

SIMD processor architectures for SDR can be categorized into short SIMD architectures and wide SIMD architectures. Most of the SIMD processors from both classes support long instruction word (LIW) execution (see chapter 3.1.3), enabling concurrent memory access and arithmetic operations. Short SIMD architectures support SIMD data paths with few parallel lanes (typically four lanes). Examples are Sandbridge's Sandblaster SB3011 [Gl06], Icera's DXP [Kno05], Infineon's MuSIC archi-

tecture [GRS07] and Linköping University's single instruction stream, multiple task (SIMT) architecture [Nil07, NTL09].
Wide SIMD architectures utilize 16 or more parallel data lanes. A higher degree of SIMD parallelism leads to better energy efficiency, but higher levels of data parallelism are required to utilize all data lanes effectively. This book analyses data parallelism in SDR algorithms for wide SIMD architectures. Hence, relevant wide SIMD architectures are discussed in more detail in section 2.3. The CEVA-XC [CEV09] architecture is a wide SIMD processor architecture, which supports VLIW execution of multiple parallel 256-bit SIMD data paths. Yet, too little information on CEVA-XC is available for a detailed analysis.

2.2 Basic principles of SIMD processing

The term single instruction stream, multiple data stream (SIMD) processing has been introduced by Flynn in his taxonomy of very high-speed computers [Fly66, Fly72]. Flynn classified processor architectures based on the parallel processing of instruction and data streams. A SIMD processor architecture processes multiple data streams in parallel, yet each data stream performs the same operation, as there is only one instruction stream. SIMD processor architectures for SDR support SIMD processing by vector operations on fixed-length data vectors.

2.2.1 SIMD vector processing

SIMD vector operations process all *vector elements* independently and in parallel. Vector *data lanes* contain the processing units associated to vector element positions. The number of vector elements is the *vector length* or *SIMD width*. Figure 2.2 shows three examples of vector operations, which demonstrate the principle of vector processing.

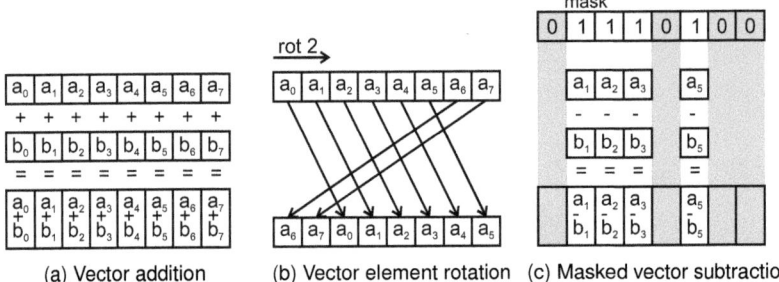

(a) Vector addition (b) Vector element rotation (c) Masked vector subtraction

Figure 2.2: Examples for SIMD vector operations

The vector addition in figure 2.2a adds elements from one vector to elements from another vector. All data lanes operate independently of each other, yet they all execute the same instruction. Furthermore, only elements at the same position in a vector can be processed together, it is e.g. not possible to add element b_3 to element a_1 in a vector operation.

If vector elements are needed at different positions, they have to be permuted on a *permutation network*. Figure 2.2b shows a cyclic rotation of vector elements as an example for a vector element permutation.

Figure 2.2c demonstrates a third technique that is necessary for efficient SIMD vector processing. If an operation shall only be performed on some of the elements in a vector, the remaining element positions can be excluded from the computation by *masking*. Masking temporarily disables vector lanes based on the values of a binary vector mask.

Figure 2.3 illustrates the structure of a typical SIMD processor architecture with four parallel lanes. The lanes contain one or multiple processing units (in this case an ALU) for the processing of one vector element and a local register file that stores the elements of different vectors. All lanes have the same structure and the processing units in different lanes execute the same instructions in parallel. Memory access is also done in parallel on a load/store unit. Element permutations are performed on the vector permutation unit, which contains a permutation network — in this case, based on an inverse butterfly network topology (see chapter 3.1.5).

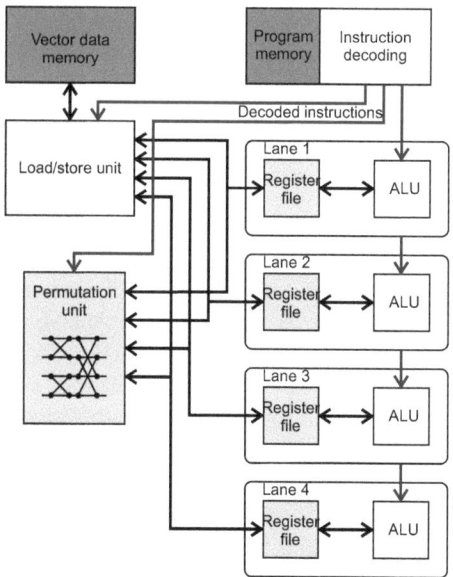

Figure 2.3: Example SIMD processor architecture with four lanes

2.2.2 Advantages and disadvantages of SIMD processing

The main reason for using SIMD processor architectures in SDR systems is their high energy efficiency. Scalar processors spend significant amounts of the total power on the decoding of instructions and other control tasks. SIMD processors execute one instruction on many parallel data paths; hence, the relative amount of power spent on control tasks is reduced. Wider SIMD vectors lead to higher energy efficiency. Furthermore, parallel processing enables lowering the clock frequency, which also reduces power consumption.

Due to the regular structure of the SIMD data path (see figure 2.3), SIMD processor architectures scale very well. Only the number — not the complexity — of data lanes increases if the vector length is raised. The only exception is the vector permutation network, whose complexity depends on the number of vector elements (see chapter 3.1.5).

The main drawback of SIMD processor architectures is that they only reach high energy efficiency if the vector data lanes can be effectively utilized. The utilization of the data lanes is limited by available *data parallelism*, i. e. the number of data values that can be processed in parallel, and the restriction by the required alignment on vectors. Due to the restricted alignment, some algorithms require complex permutation stages for reordering vector elements. A further drawback is the lack of efficient software development tools that can perform the vectorization of algorithms. Hence, software development for SIMD processors is time consuming and requires knowledge about the processor architecture.

Data parallelism in baseband processing

Many baseband processing algorithms, such as FIR filtering, FFT, and correlation, offer plenty data parallelism, with very little data dependent control flow. Most algorithms can be represented by matrix-vector or vector-vector operations. Hence, SIMD processing is a natural fit for SDR.

The amount of available data parallelism and the overhead for performing data alignment operations (e. g. vector permutations, memory access) depend on the algorithms and on architecture parameters. The overhead for vector permutation operations and memory access can be reduced by enabling parallel processing of operations on distinct processing units, e. g. through LIW execution.

The goal of this book is analyzing the mapping of three major 4G baseband algorithms on wide SIMD processor architectures with LIW support. The dependence between algorithm parameters and performance for different SIMD widths is evaluated. The focus is on determining how performance and energy efficiency scale with the SIMD width.

2.3 Wide SIMD processor architectures and research on the scalability of SIMD processing

In the following, relevant academical and commercial wide SIMD processor architectures for SDR are surveyed. The discussed processor architectures are the Embedded Vector Processor (EVP) (section 2.3.1), the Sandblaster SB3500 processor (section 2.3.2), the Signal-processing On-Demand

Chapter 2 Overview of software defined radio principles and architectures

Architecture (SODA) (section 2.3.3), and the Ardbeg processor (section 2.3.4). Architectural features of these processor architectures are compared in table 2.1. Here, the performance is measured in giga MAC operations per second (GMAC/s) on 16-bit values.[1] Most of the processor architectures support LIW execution and permutation operations on SIMD permutation networks. These aspects of the SIMD processor design are discussed in detail in chapter 3, sections 3.1.3 and 3.1.5.
In section 2.3.5, single instruction, multiple disjoint data (SIM_dD) architectures are discussed. Section 2.3.6 reviews related research on the scalability of SIMD processing for SDR.

2.3.1 The Embedded Vector Processor

The Embedded Vector Processor (EVP) is a SIMD vector processor architecture that is commercially available from ST-Ericsson [SE09] (formerly NXP Semiconductors). The processor architecture has been designed for SDR systems, focusing on the modem stage of the baseband processing [vHM+04, vHM+05]. The first instance of the EVP, the VD32040 processor is already in mass-production in multi-mode TD-SCDMA handsets [SVPG+10]. A block diagram of the processor architecture is displayed in figure 2.4.

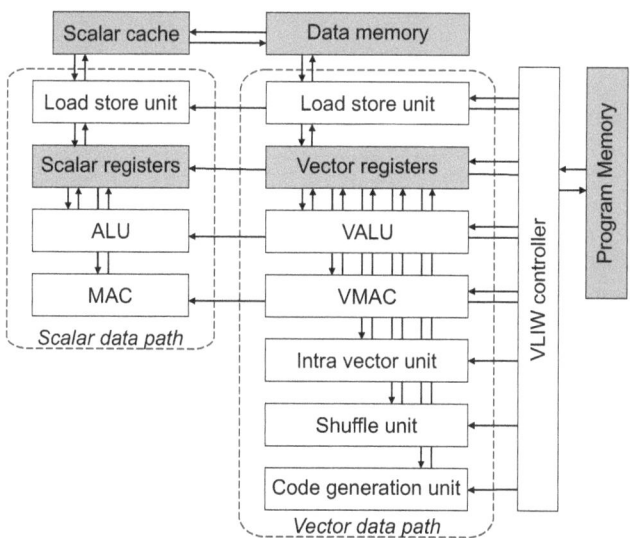

Figure 2.4: Block diagram of the EVP architecture

The EVP supports vector operations on 256-bit data vectors. Data vectors can be interpreted as 8 32-bit elements, 16 16-bit elements, or 32 8-bit elements. Yet, some instructions (e. g. multiplication, MAC operation) only support 16-bit values. Complex multiplication and MAC operations are also supported.

---
[1] For SODA and Ardbeg, performance is reported in giga multiplications per second (GMul./s).

Table 2.1: Overview of commercial and academical SIMD processors for SDR

Architecture	EVP	Sandblaster SB3500	SODA	Ardbeg
Processor core architecture				
Technology	45 nm	65 nm	$0.18\,\mu m / 0.13\,\mu m$	90 nm
Multi-core	1 core	3 cores, 4 threads per core	4 cores	2 cores
LIW slots	10 (6 vector)	3 (vector + memory + scalar or branch)	1 vector	2 vector
Max. frequency	320 MHz	600 MHz	400 MHz / 300 MHz	350 MHz
Local memory (data+program)	128 KB + 128 KB	256 KB + 32 KB per core	8 KB + 4 KB	32 Kb – 128 KB
Performance	4.8 GMAC/s (300 MHz)	28.8 GMAC/s	19.2 GMul./s (300 MHz)	22.4 GMul./s
Power/Energy	<0.5 mW/MHz	300 mW @ 500 MHz	$\approx$3 W $(0.18\,\mu m)$ 210 mW $(0.13\,\mu m)$	$\approx$180 mW (W-CDMA 2 Mbps)
Area	$3.2\,mm^2$	—	$26.6\,mm^2$ $(0.18\,\mu m)$ $10\,mm^2$ $(0.13\,\mu m)$	—
SIMD architecture				
SIMD width	256 bit	256 bit	512 bit	512 bit
Data precision	8/16/32 bit	16/32 bit	16 bit	8/16/32 bit (block float. point)
Complex mult.	✓	✓	N/A	N/A
SIMD reg. file	16 registers	8 registers	16 registers	15 registers
Permutation network	crossbar	—	single-stage perfect shuffle	double-vector banyan network

The processor architecture consists of a scalar and a vector data path. The scalar data path contains an arithmetic logic unit (ALU), a load store unit for scalar memory access (LSU) and a multiply-accumulate unit (MAC). Furthermore, scalar values can be broadcasted to vectors. The scalar program control unit (PCU, not in figure 2.4) performs branch operations and zero-overhead loops (ZOLs). The vector data path contains a vector ALU (VALU), a vector MAC unit (VMAC), a vector LSU (VLSU) and three special vector processing units: The vector shuffle unit (VSHU) performs arbitrary vector element permutations on one data vector. The intra vector unit (IVU) supports minimum/maximum search on elements of a vector and accumulation of elements in a vector. The code generation unit (CGU) is an application-specific accelerator for scrambling and channelization codes (e. g. for W-CDMA).

Next to processing data vectors in parallel, the EVP also supports very long instruction word (VLIW) execution: Operations on each of the vector and scalar processing units can execute in parallel in one clock cycle.

The processor is designed for a working frequency of 300 MHz (maximum 320 MHz in a low-power 45 nm process) and achieves 4.8 GMAC/s in the SIMD data path and up to 30 GOPS (Giga operations per second) on 16-bit data at this clock frequency [SE09]. For a typical memory configuration (128 KB program memory, 128 KB data memory) and typical applications, the newest instance of the EVP, the VD32041, requires less than 0.5 mW/MHz energy and $3.2\,\text{mm}^2$ area ($1.3\,\text{mm}^2$ for the processor core without memories).

The architectural features of the scalable SIMD processor architecture in chapter 3 are based on an extensive analysis of SDR applications on the EVP.

2.3.2 The Sandblaster SB3500 architecture

The Sandblaster SB3500 architecture is a low-power multi-core processor architecture developed by Sandbridge Technologies [MG08, SMN+09, SVPG+10]. The multi-core processor consists of an ARM processor core and three SBX processor cores. The SBX cores support 256-bit SIMD vector operations, including complex multiplication, reduction operations (e. g. sum of vector elements), permutation operations (rotation, shifting) and special operations for Galois field arithmetic and Viterbi decoding.

Each SBX executes four sequential hardware threads. If the processor is clocked at the maximum clock frequency of 600 MHz, this corresponds to four threads running at 150 MHz. The sequential processing of threads enables an efficient pipelining of execution stages, which hides the instruction latency from the programmer and reduces the power consumption. Each SBX core also supports long instruction word (LIW) execution; three operations are issued per clock cycle, including up to one vector operation and one memory access operation.

The SB3500 is supported by a complete parallelizing tool chain. The compiler includes algorithms for the vectorization of loops, the detection of saturation arithmetic, and multi-threading.

The SB3500 processor architecture achieves a peak performance of 9.6 GMAC/s per SBX core and is fabricated in 65 nm technology. The typical power consumption of one SBX core at 500 MHz is 100 mW [San09].

2.3.3 The Signal-processing On-Demand Architecture

The Signal-processing On-Demand Architecture (SODA) is a multi-core SIMD processor architecture for SDR [LLW+06, LLW+07, Lin08, WSL+07], which was developed at the University of Michigan at Ann Arbor. The original SODA consists of four processing elements (PEs) with local memories, a global memory, and an ARM control processor (see figure 2.5). The PEs support SIMD operations on 512-bit data vectors.

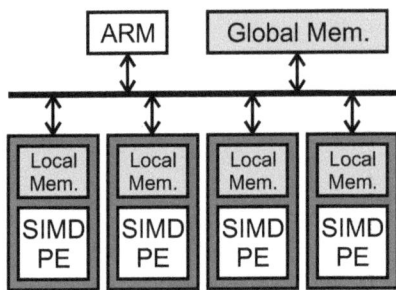

Figure 2.5: Block diagram of SODA

Each PE has three parallel pipelines for address generation, scalar, and SIMD operations. Each pipeline executes one instruction per clock cycle. The SIMD pipeline consists of 32 lanes with 16-bit ALUs and 16-bit multipliers (a multiplication requires two clock cycles). A so-called SIMD shuffle network (SSN) performs vector element permutations using a single-stage combined perfect shuffle exchange, inverse perfect shuffle exchange network with a feedback path [LLW+06], which enables performing complex permutations by multiple permutation iterations on the single-stage network. Figure 2.6 shows the topology of perfect shuffle exchange and inverse perfect shuffle exchange networks for eight elements in a vector.

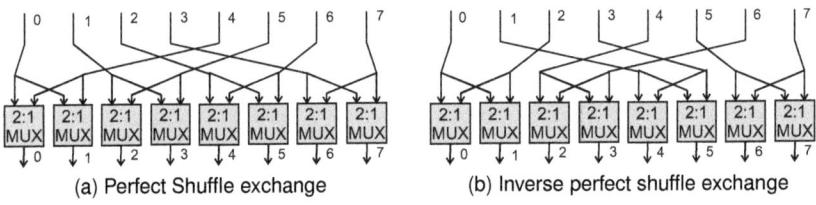

(a) Perfect Shuffle exchange   (b) Inverse perfect shuffle exchange

Figure 2.6: Topologies of perfect shuffle exchange and inverse perfect shuffle exchange networks

SODA has been synthesized in 0.18 μm technology for a maximum clock frequency of 400 MHz. The total power consumption for W-CDMA (2 Mbps) is 2.95 W, SODA requires $26.6\,\text{mm}^2$ for PEs, ARM processor, bus system, and memories. Recently, the processor architecture has been resynthesized in $0.13\,\mu$m technology [LCM09] for a clock frequency of 300 MHz. In $0.13\,\mu$m technology, the architecture requires 210 mW for W-CDMA (2 Mbps) and occupies $10\,\text{mm}^2$ chip area. The processor architecture achieves 19.2 GMul./s at 300 MHz.

SODA-II

SODA-II is a follow-up architecture of SODA, which tries to address the shortcomings of the original SODA by optimizing the data paths [LCM09]. The main shortcomings of the original SODA are its high register file power consumption (between 27 and 37 percent of the total power consumption [LLW+06]) and the overhead for data alignment operations. SODA-II applies three techniques to improve SODA: *operation chaining, pipelining of vector operations,* and *staggered memory access and multi-cycling.*

Operation chaining describes the concatenation of SIMD operations in one instruction. SODA-II's SIMD computation units comprise two parallel ALUs/multipliers, followed by a third ALU. Operations, such as performing two multiplications and accumulating the results can be performed in one instruction. The concatenation of instructions lessens register file accesses and reduces the cycle count. A simplified block diagram of a SIMD computation unit is shown in figure 2.7.

Pipelining of vector operations is done to reduce the overhead for non-computational operations, such as memory access and vector permutations. A conceptual view of SODA-II's vector pipeline is shown in figure 2.8. The pipeline comprises five stages: memory read access, vector alignment, vector computation, vector alignment or reduction, and write back to memory. The vector alignment is realized on the SIMD permutation network or by simply writing vectors to vector registers.

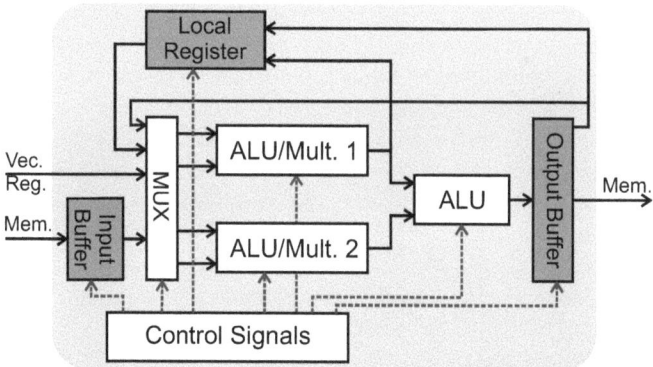

Figure 2.7: Simplified block diagram of a SIMD computation unit in SODA-II

Staggered memory access and multi-cycling are done to avoid permutations for complex data alignments, such as required for the FFT or Viterbi decoding. Complex data alignment is implemented by reading one value at a time (denoted as staggered memory access); the processing on computation units for different vector elements is initialized sequentially. A high utilization of the SIMD computation units with staggered memory access can only be achieved if the computation units are busy for multiple cycles (denoted as multi-cycling). Therefore, the computation units are realized as

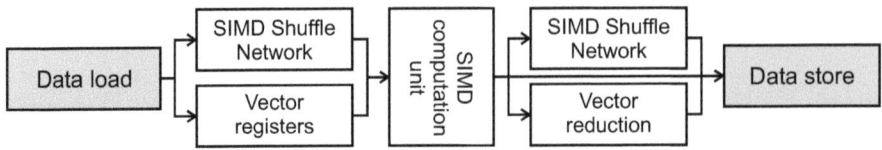

Figure 2.8: Conceptual view of the vector pipeline in SODA-II

small microprocessors, which support complex multi-cycle operations that iterate on the input data for multiple cycles and single-cycle operations, such as the chained operations mentioned above. Like SODA, SODA-II is also realized as a multi-core architecture with four 512-bit SIMD processors. SODA-II has been synthesized in a 0.13 μm process for a clock frequency of 300 MHz [LCM09]. The processor architecture requires $11\,\text{mm}^2$ chip area. W-CDMA (2 Mbps) requires two of the four SIMD processor at a power consumption of approximately 120 mW. The original SODA requires all four processors to achieve 2 Mbps throughput and consumes 210 mW power. Performance results for 4G algorithms have not been obtained.

2.3.4 The Ardbeg architecture based on SODA

The Ardbeg architecture is a commercial processor architecture, which has been developed based on a redesign of SODA for 90 nm technology [WLS+08a, Lin08]. The processor architecture is part of ARM spin-out Cognovo's software defined modem platform. The Ardbeg architecture contains two Ardbeg PEs. Figure 2.9 shows a high-level block diagram of the data path of an Ardbeg PE. SODA has been extended with limited LIW support for SIMD operations and an accelerator for turbo decoding (ASIC). Furthermore, the permutation network has been redesigned and SIMD processing units have been optimized. The SIMD width (512 bits) has not been changed.

LIW support has been implemented, as the utilization of processing units in SODA is poor. For example, the SIMD ALU is only active for 30 percent of the total time for W-CDMA and IEEE 802.11a [WLS+08a]. LIW support can improve the resource utilization by e.g. enabling to perform memory access in parallel to arithmetic operations. In the Ardbeg architecture, LIW processing is limited to at most two parallel SIMD operations, and only some combinations of functional units are supported (e.g. memory access and multiplication or multiplication and ALU operation).

Turbo decoding has been offloaded to an accelerator, because of its poor performance on SODA: The turbo decoder for W-CDMA, at a data rate of 2 Mbps, occupies one complete SODA PE. The same performance can be achieved on an ASIC with approximately 5 times lower power consumption.

SODA utilizes a single-stage permutation network, which can only perform a very limited set of permutations in one step. Multi-stage interconnect networks (MINs) have not been considered, because the delay in $0.18\,\mu$m technology was too long. Ardbeg has been designed for 90 nm and implemented with a banyan network with seven stages and a width of two vectors. The banyan network is a MIN and can perform permutations on pairs of vectors in one operation (see chapter 3.1.5).

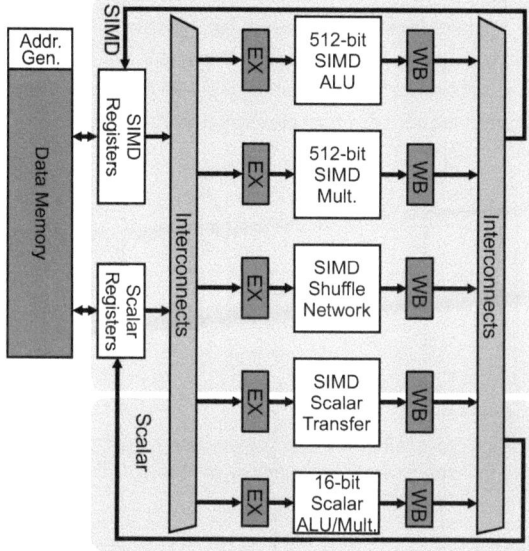

Figure 2.9: Block diagram of the SIMD and scalar data paths of an Ardbeg PE

As a last optimization, the SIMD processing units have been optimized. All units support 8-, 16-, and 32-bit data types instead of only 16-bit data types. Furthermore, block floating point support has been added. In a floating point representation, each value has its own mantissa and exponent. In a block floating point representation, all elements in a vector share the same exponent, which reduces the overhead for floating point processing. Moreover, a single-stage permutation network has been added to the SIMD ALU, which enables to perform permutations (e. g. for an FFT) and addition or subtraction in one operation. The multipliers have been redesigned for single-cycle latency.
The various optimizations lead to a speedup between $1.5\times$ and $7\times$ over SODA for W-CDMA, IEEE 802.11a, and DVB-H/DVB-T [WLS+08a]. The Ardbeg architecture consumes approximately 180 mW of power for W-CDMA at 2 Mbps.

2.3.5 Processor architectures based on SIM$_d$D processing

Single instruction, multiple disjoint data (SIM$_d$D) processing [HM03] describes a processor architecture that enables SIMD units to operate on any data, without alignment restrictions by forcing operations on data vectors. The register file complexity of a SIM$_d$D architecture grows with the SIMD width as the register file complexity of a VLIW architecture grows with the number of parallel instruction slots. Therefore, SIM$_d$D processing is not practical for wide SIMD architectures.
Indirect SIM$_d$D architectures still operate on aligned data vectors in registers, yet allow flexible memory access by providing vector pointers. The basic idea is illustrated by figure 2.10: A vector pointer

is provided; each component of the vector pointer determines the memory location of one vector element or one vector segment (if the vector pointer has fewer elements than the data vectors) for memory access operations. During memory access, the vector elements/segments are processed independently. Data that has been read from memory is stored in data vectors (in a register file). The processing of data vectors does not differ from classical SIMD architectures.

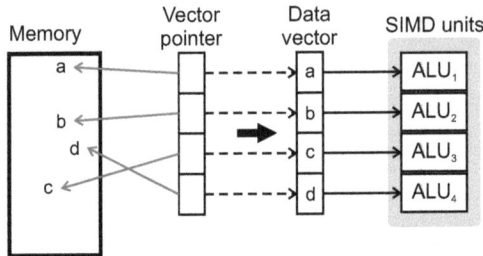

Figure 2.10: Indirect SIM$_d$D processing: A vector pointer determines memory locations for each vector element independently during memory access operations

IBM's eLite DSP architecture

The concept of SIM$_d$D processing has been invented at IBM research and implemented in the eLite DSP architecture [MZS+03, DMW03, DM03]. eLite supports 64-bit SIMD vectors (four 16-bit elements) and LIW execution. The vector data path consists of a vector pointer unit, a vector element unit, and a vector accumulator unit. The vector pointer unit performs vector pointer operations for addressing registers in the vector element register file. The vector element unit performs arithmetical operations on the data from the vector element register file. The vector accumulator unit is utilized for accumulation or reduction operations on 40-bit values. Every vector instruction specifies one or two vector pointer registers, which in turn point to the element registers.

eLite has been implemented in $0.13\,\mu$m technology and achieves 2 GMAC/s at a clock frequency of 500 MHz with a power consumption of 300 mW. At 250 MHz eLite only consumes 50 mW of power. While eLite is not an alternative for SDR processing, the concept of SIM$_d$D — and especially indirect SIM$_d$D — is interesting for wide SIMD architectures, as discussed in chapter 7.3.1.

The AnySP architecture

The AnySP architecture has been developed at the University of Michigan at Ann Arbor [WSM+09, WSM+10, WMMC10] and is the first indirect SIM$_d$D processor architecture for SDR. Like SODA, AnySP is a multi-processor architecture with four processing elements (PEs). Each AnySP PE is a 1024-bit SIMD processor (64 16-bit lanes) with a configurable SIMD data path that supports three modes of execution: wide SIMD execution with 64 lanes, eight parallel threads with eight lanes, and SIMD execution with 32 lanes and operation chaining (see section 2.3.3). A block diagram of an AnySP PE is displayed in figure 2.11.

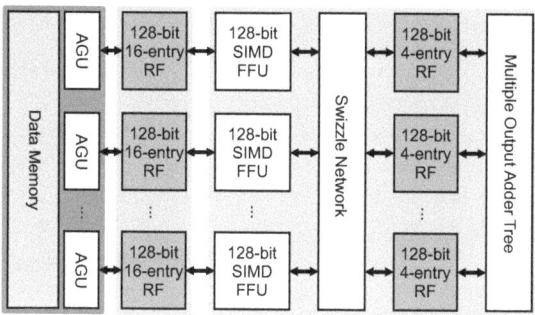

Figure 2.11: Block diagram of an AnySP processing element

Execution with 32 lanes and operation chaining is realized by using two neighboring 16-bit processing units in consecutive cycles without write back to the register file. Hence, the vector length is reduced for support of more complex instructions. Neighboring units that can be combined are denoted as flexible functional units (FFU).

Eight parallel threads on 128-bit vectors are supported by indirect SIM_dD processing, each thread executes the same operations, only memory access locations differ. Memory access is realized by providing eight independent address generation units (AGUs), one for each 128-bit SIMD thread. Furthermore, accumulation of vector elements for independent threads is supported by implementing a multiple-output adder tree.

The architecture has been further optimized by providing two SIMD register files (RFs): one large register file (16 entries) and one small register file (4 entries) for values with a short lifetime. The partitioning of the register file into two register files reduces the power consumption. Furthermore, a crossbar network based on SRAM cells for cross points, called swizzle network, is implemented. The SRAM-based network requires less power than a multiplexer-based network.

AnySP has been synthesized in 90 nm technology for a target frequency of 300 MHz. Performance and power consumption have been measured based on a mix of H.264 decoding (4CIF, 30 fps) and wireless communication algorithms (1024-point FFT, 2×2 STBC and decoding of quasi-cyclic LDPC codes). The wireless communication algorithms achieve a throughput of 100 Mbps. The total power consumption is 1.347 W, the chip area is $25.17 \, mm^2$.

2.3.6 Research on the scalability of SIMD processing for SDR

While SIMD processing is a technique that has been thoroughly investigated, few research results have been published that discuss the scalability of wide SIMD processor architectures for SDR algorithms. [WLS+08a] describes the development of the Ardbeg architecture from SODA. A limited analysis of different SIMD widths and permutation networks has been done before selecting the final SIMD width for Ardbeg. [WLS+08b] investigates the scalability of SODA for four baseband algorithms.

SIMD scalability analysis during the Ardbeg development

[WLS+08a] includes a SIMD width analysis for the Ardbeg processor in 90 nm technology. Ardbeg has been synthesized for SIMD widths ranging from 128 to 1024 bit (8 to 64 16-bit lanes). The synthesized permutation network is not mentioned. Energy consumption, delay, and area have been measured for a mix of 3G baseband algorithms, including FIR filtering, FFT, W-CDMA searcher (based on auto-correlation), and Viterbi decoding. Algorithm parameters (e. g. filter length, FFT size) are not mentioned. Results are reported as an average over all algorithms and are normalized to the 128-bit SIMD processor. The normalized delay figures show approximately linear speedup.[2] The energy consumption decreases with an increasing SIMD width (60 percent for 64 lanes). The area more than doubles with a doubling of the SIMD vector length (approximately $10\times$ for 64 lanes). Due to the significant increase in area, the SIMD width has been set to 32 lanes and not 64 lanes.

After fixing the SIMD width, four different permutation networks have been implemented and synthesized and normalized energy and energy-delay-product have been measured for 64-point and 2048-point radix-2 and radix-4 FFTs and a Viterbi decoder for constraint length 9. The implemented permutation networks are SODA's single stage perfect shuffle exchange / inverse perfect shuffle exchange network with a width of one vector (enabling permutations on one input vector), the same network topology with a width of two vectors (enabling permutations on pairs of vectors), a banyan network (multistage interconnect network - MIN) on two vectors, and a crossbar network on two vectors. More information on permutation network topologies can be found in chapter 3.1.5. The analysis shows that the algorithm implementations with permutation networks with a width of two vectors consume less energy and have a better delay than the implementations with a single-vector network. The banyan network and the crossbar network achieve similar results, except for the 64-point radix-2 FFT, which has a much higher energy consumption using the crossbar network. The double-vector perfect shuffle exchange / inverse perfect shuffle exchange network achieves the best results for the 64-point radix-2 FFT, as the FFT algorithm is optimized for this network architecture. The banyan network and the crossbar network attain the best results for all remaining algorithms. Based on the analysis, Ardbeg has been realized with a double-vector banyan network.

SIMD scalability analysis based on SODA

The SIMD scalability analysis in [WLS+08b] based on SODA considers four SDR algorithms for MIMO-OFDM for SIMD widths ranging from 512 bit (32 16-bit lanes) to 4096 bit (256 16-bit lanes). The implemented algorithms are a 1024-point radix-2 FFT, space time block coding (STBC) based on Alamouti for a 2×2 MIMO system [Ala98, Bau01], the vertical Bell laboratories layered space-time (V-BLAST) detection algorithm for 4×4 MIMO [Fos96, WFGV98], and a decoder for a WiMAX LDPC code ($z = 96$, $R = 5/6$) [IEE09b, SMZC07].

First, available data parallelism and workload have been analyzed; the results are summarized in table 2.2. According to Woh et al., the SIMD width should be increased to be as large as the FFT size N_{DFT} for the maximum performance. STBC and V-BLAST both operate on small vectors, with one

---
[2]The term *linear speedup* or *ideal speedup* means that a doubling of the SIMD width leads to a doubling of the performance.

vector per OFDM sub-carrier. As sub-carriers are orthogonal to each other and can be processed in parallel, data parallelism is only limited by the number of data carriers in an OFDM symbol, which here is assumed the FFT size. The LDPC decoder operates on $z \times z$ sub-matrices of the LDPC matrix; hence, at most z elements can be processed in parallel.

Table 2.2: Analysis of data parallelism [WLS+08b]

Algorithm	Overhead Workload [%]	Scalar Workload [%]	SIMD Workload [%]	Maximum vector parallelism
FFT/IFFT	61	5	34	$N_{DFT} = 1024$
2×2 STBC	14	5	81	$4 \cdot N_{DFT}$
4×4 V-BLAST	24	6	70	$4 \cdot N_{DFT}$
LDPC	3	18	49	$z = 96$

The workload results categorize the workload on the SIMD processor into scalar workload, workload for computational SIMD operations on the ALU, multiplier, or shifter (denoted as SIMD workload), and overhead workload for memory access and vector permutations. The workload results show a high utilization of the computational SIMD units for STBC and V-BLAST, the FFT is dominated by overhead workload.

The speedup and energy consumption have been measured and normalized to the results for 32 16-bit lanes. Normalized speedup results show linear speedup for STBC and slightly less than linear speedup for FFT. The V-BLAST implementation apparently requires more scalar operations for wider SIMD widths; hence, the speedup increases slowly (approximately $5.5\times$ for 256 lanes). The speedup for LDPC decoding also increases slowly and does not increase at all if the SIMD width is increased from 128 to 256 lanes, as at most 96 elements can be processed in parallel. The maximum speedup is 3.0 for 128 or more parallel lanes. If linear or close to linear speedup can be attained, the energy consumption stays almost constant. The LDPC decoder requires more energy on wider SIMD architectures, because most of the SIMD lanes perform useless computations.

Differences to the present work

The analysis of the scalability of SIMD processing in this book differs from the work in [WLS+08b, WLS+08a] concerning the analyzed algorithms and the considered SIMD architecture.
SODA supports only one vector operation per clock cycle, while Ardbeg supports a restricted LIW instruction format with at most two parallel vector operations. Therefore, much of the processing time is spent either on scalar or memory access and vector alignment operations, as can be seen in table 2.2. This paper proposes a SIMD processor architecture with LIW support. Parallel processing of computational SIMD operations (e.g. addition, multiplication) and memory access and/or vector permutation operations increases the performance of the SIMD architecture, as overhead operations can be hidden by LIW execution. One prominent example is the FFT with an overhead workload of 61 percent on SODA. On the proposed SIMD architecture, the overhead operations can be completely or mostly performed in parallel to useful computational operations (see chapter 4.6.3).

The SIMD processor architecture has also been implemented with four different permutation network configurations (see chapter 3.1.5), enabling to perform a systematic analysis of the complexity of permutations for different SIMD widths.

From an algorithm perspective, the analysis in this book considers different algorithm parameters, a more recent MIMO detection algorithm, and — in part — achieves different results than the work in [WLS+08b, WLS+08a].

Woh et al. only implemented one FFT size ($N_{DFT} = 1024$) and one LDPC code ($z = 96$, $R = 5/6$). Hence, the influence of algorithm parameters could not be investigated. This paper analyzes different FFT sizes, including mixed-radix FFT sizes and LDPC codes. The results show that the scalability indeed depends on algorithm parameters.

In [WLS+08b], 2×2 STBC and 4×4 V-BLAST are implemented as examples for MIMO algorithms. STBC is an approach that increases the signal quality at the receiver by sending a signal on multiple transmit antennas in a space-time code [Ala98]. V-BLAST is a detection algorithm for spatial multiplexing, i. e. multiple data streams are transmitted in parallel [WFGV98]. This paper analyzes 4×4 sphere decoding, which is a class of detection algorithms for spatial multiplexing. V-BLAST has a lower computational complexity than sphere decoding, but has a poor BER performance, as it does not exploit the full MIMO diversity [BBW+05]. Sphere decoding algorithms achieve a BER performance close to the optimum maximum likelihood (ML) solution. Therefore, sphere decoding is a better choice for 4G systems — and more challenging due to the greater computational complexity.

Woh et al. also arrive at different conclusions concerning the scalability of FFT and LDPC decoding than this paper. [WLS+08b] claims that the SIMD width should be increased to the FFT size, while the LDPC decoder can process at most z elements in parallel. In chapter 4, it is shown that the FFT size should be at least twice the SIMD width for radix-2 FFTs. Chapter 6, which describes the LDPC decoder implementation, shows that LDPC decoding may also be efficiently done for SIMD widths greater than z, yet a different implementation is required than for SIMD widths less than or equal to z.

2.4 Key algorithms for future 4G SDR systems

The requirements of 4G wireless systems are defined by the ITU [Rep08] under the identifier International Mobile Telecommunications-Advanced (IMT-Advanced). IMT-Advanced defines a target data rate of 100 Mbps for high and 1 Gbps for low mobility environments. The key technologies that may enable these data rates are orthogonal frequency division multiple access (OFDM-A), multiple-input, multiple-output (MIMO) transmission and forward error correction (FEC) coding with turbo codes and LDPC codes.

The future Long Term Evolution-Advanced (LTE-Advanced) standard [Tec09a] is the first candidate for an IMT-Advanced standard and targets a downlink peak data rate of 1 Gbps and an uplink peak data rate of 500 Mbps. LTE-Advanced is based on the pre-4G Long Term Evolution (LTE) standard [Tec09b], which supports up to 4×4 MIMO and uses OFDM-A for downlink and single-carrier frequency division multiple access (SC-FDMA) for uplink transmission. FEC coding is done by convolutional codes and Turbo codes. The other major pre-4G wireless transmission standard, Worldwide Interoperability for Microwave Access (WiMAX, IEEE 802.16) [IEE09b], is also based on MIMO-OFDM and optionally supports LDPC codes. Hence, a MIMO-OFDM system with turbo or LDPC codes for channel coding can be identified as the probable foundation for the physical layer processing of future IMT-Advanced compliant transmission standards. Below, the most important tasks in a MIMO-OFDM system are established.

2.4.1 MIMO-OFDM system model

OFDM is a block modulation scheme that converts a frequency-selective channel into many frequency flat sub-carriers [SBM+04, Bö6]. The sub-carriers are orthogonal in time domain, yet overlap in frequency domain, which leads to a good bandwidth efficiency. OFDM is realized by an IFFT at the transmitter and an FFT at the receiver. In a MIMO-OFDM system, OFDM block modulation is used for multiple transmit antennas and the receiver applies multiple receive antennas. A block diagram of a MIMO-OFDM system is displayed in figure 2.12.

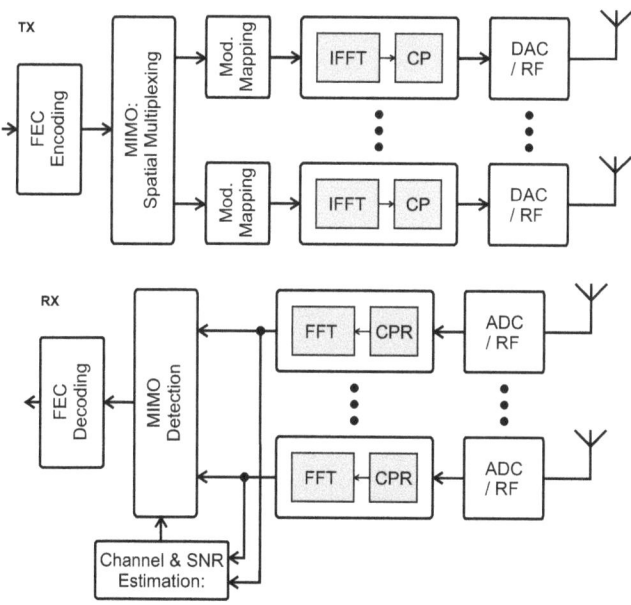

Figure 2.12: Block diagram of transmitter and receiver in a MIMO-OFDM system

From a complexity point of view, the receiver side is the most demanding part of the physical layer in a MIMO-OFDM system. The major tasks that have to be performed at the receiver are the OFDM demodulation by an FFT, the detection of the most likely transmitted MIMO symbols (MIMO detection), the decoding of the FEC code (e. g. a turbo or LDPC code), time synchronization, frequency offset estimation and correction, and the estimation of channel parameters (i.e. channel matrix and SNR) [SBM+04].

Time synchronization, frequency offset estimation and correction, and channel estimation are usually performed using a preamble consisting of one or several training sequences [SBM+04]. Once the necessary parameters have been computed, the receiver tracks changes for the duration of a frame, e.g. using pilot signals. None of these tasks depends on the incoming data.

OFDM demodulation, MIMO detection, and FEC decoding are tasks that have to be performed continuously on the incoming data, with the throughput requirements defined by the data rate of the MIMO-OFDM system. Hence, the computational complexity of the receiver is mostly defined by these three tasks. Therefore, OFDM-based block (de-)modulation, MIMO detection, and LDPC decoding have been selected to assess the potential performance — and the limitations for SIMD processing — of MIMO-OFDM systems on an SIMD-based SDR processor platform.

LDPC codes and turbo codes both can achieve error rates close to then Shannon limit. Yet, LDPC codes have an asymptotically better performance than turbo codes and enable trade-offs between decoding complexity and performance [RSU01, RU01]. Furthermore, LDPC decoding algorithms are well suited for parallel processing. Therefore, LDPC decoding has been investigated instead of turbo decoding.

MIMO symbol detection is performed based on sphere decoding, because sphere decoding algorithms offers an excellent BER performance (close to the optimum maximum likelihood solution). Furthermore, the high computational complexity of sphere decoding is a challenge for any baseband processing architecture.

Chapter 3
Scalable SIMD processor architecture

This chapter focuses on the development of a scalable SIMD processor architecture for SDR applications. The scalable SIMD processor architecture is described and design decisions are explained. Furthermore, the modeling in LISA and the methodology for evaluating the scalability of the architecture are explained. The scalable architecture has been synthesized and simulated for four different SIMD vector widths — ranging from 128 bits to 1024 bits per vector — and four different permutation network configurations.

In section 3.1, design decisions for the scalable SIMD processor architecture are explained. The architecture is developed based on an analysis of baseband algorithms that have been implemented on the EVP [vHM$^+$04, vHM$^+$05, SVPG$^+$10]. Following the description of the processor architecture, section 3.2 explains the processor modeling using the LISA language [HNBM01, Hof02, Lö4, PHZM99, Pee02]. After briefly introducing LISA, modeling issues for scalable SIMD models are described and a solution based on the GNU M4 [SPVB08] macro language is proposed. The section concludes with a discussion of the effectiveness of LISA for the development of large SIMD processors. The following section (section 3.3) briefly discusses an alternative for LIW SIMD architectures based on vertical-horizontal vector operations [GZYC86]. Section 3.4 explains the methodology used for analyzing area, power, and performance of the different instances of the scalable SIMD processor architecture. The used tools and technologies are described and limitations of the methodology are explained.

3.1 Development of the SIMD processor architecture based on algorithm requirements

The development of any processor architecture requires many design decisions. Instruction set and data type support have to be defined, register files must be dimensioned, and other features have to be defined as well. These decisions are not arbitrary and require careful consideration of the hardware complexity and, especially, the demands of the algorithms that shall be mapped on the processor architecture.

Hence, as a first step towards the development of a scalable SIMD processor architecture for SDR, the requirements of typical baseband algorithms have been identified. For this purpose, baseband algorithms that have been implemented on the EVP during a research project in cooperation with Nokia Siemens Networks were analyzed. Aspects such as typical word lengths, data types, useful instructions, conditional execution of operations, instruction level parallelism (ILP), register files, and support for permutation operations have all been considered. The implemented baseband algorithms that are the basis of the analysis are listed below:

- A linear minimum mean squared error (LMMSE) chip equalizer for single input, multiple output (SIMO) W-CDMA with two receive antennas and two times oversampling at the receiver [WBAHS08b, WBAHS09]

- A spreader for HSDPA, which comprises the modulation mapping, coding by channelization and scrambling codes, and the combining of different control and data channels [Tec07, WBAHS08b, WBAHS09]

- Matrix algorithms for MIMO OFDM systems [SM06]: QR decomposition, singular value decomposition (SVD) and the QRD-M algorithm for MIMO symbol detection

- Radix-2 and mixed-radix FFTs for single carrier frequency division multiple access (SC-FDMA) and orthogonal frequency division multiple access (OFDM-A) in LTE [WBAHS08a]

In the following, the results of the algorithm analysis are presented. Based on this analysis, the scalable SIMD processor architecture has been developed.
In section 3.1.1, word lengths and data types are evaluated. The next section evaluates instructions and conditional execution modes that are useful for the considered baseband algorithms; based on the evaluation, the instruction set of the processor architecture is defined and instructions are mapped on processing units. Sections 3.1.3 and 3.1.4 discuss instruction level parallelism and the dimensioning of the register files. In section 3.1.5, alternatives for the vector permutation network are discussed. Section 3.1.6 lists further DSP features that have been implemented and summarizes the features of the scalable SIMD processor architecture.

3.1.1 Word lengths and data types

A precision of 16 bits is adequate for the majority of considered algorithms. Some algorithms — for example the chip equalizer algorithm for SIMO W-CDMA — benefit from an increased precision for intermediate results of multiply-accumulate (MAC) operations. However, after accumulation, the result may usually be reduced to a 16 bit word length without distorting the output of the whole algorithm. In some cases, an even smaller precision than 16 bits is possible. For example, the spreading operation in HSDPA, as well as the LDPC decoder presented in chapter 6, benefit from 8-bit data types — allowing a higher memory density and twice the SIMD parallelism of an implementation with 16-bit data types.
To avoid erroneous behavior due to overflow — in case of two's complement data, a wrap around between positive and negative numbers — saturation arithmetic can be applied. Furthermore, truncation errors for multiplication operations can be reduced by added support of rounding.
Most baseband algorithms operate on complex I/Q data with mostly pure fractional data values. Hence, next to integer data types, baseband processor architectures usually support $Q.15$ fixed-point

Table 3.1: Supported basic arithmetic data types

Word length	Data format	Saturation support	Rounding support	Description
16 bits	integer	✓		default integer data type
1 bit	Boolean			Boolean data type
16 bits	$Q.15$	✓	✓	default fixed-point data type
16+16 bits	$Q.15$	✓	✓	complex-valued fixed-point data type using consecutive vector elements for imaginary and real part
40 bits	$Q8.31$	✓	✓	accumulator data type for multiplication and MAC operation
40+40 bits	$Q8.31$	✓	✓	complex-valued accumulator data type

data (one sign bit and 15 fractional bits) to accommodate these requirements. The overhead for supporting two data types (integer and fixed-point) is low, as most arithmetical operations (e.g. addition, subtraction) produce the same binary result for fixed-point and integer inputs. Only multiplication and MAC operation require different behavior for integer and fixed-point data (different output bit positions).

On a SIMD vector architecture, consecutive data elements occupy consecutive vector elements. Hence, special treatment of complex data values, where a pair of consecutive elements represents imaginary part and real part of a complex number, is expedient. Again, this only concerns the behavior of complex-valued multiplication and complex-valued MAC operation.

Based on these observations, the arithmetic data types presented in table 3.1 have been selected and implemented. All scalar and vector processing units support 16-bit data in general-purpose scalar or vector registers. The vector MAC unit (VMAC) also supports 40-bit values in a special vector accumulator register file (according to the definition in figure 3.1). Furthermore, the VMAC supports complex-valued data vectors, where imaginary and real parts are stored consecutively in pairs of vector elements. Although support of 8-bit data types is useful, a limitation to 16-bit operations, with additional support for 40-bit accumulation, significantly reduces the complexity of the SIMD processor architecture.

3.1.2 Instruction set

The instruction set of the scalable SIMD processor architecture can be divided into instructions on the scalar processing units and instructions on the vector processing units. The analysis of algorithms on the EVP shows that scalar processing is necessary for two purposes: control flow operations and the calculation of scalar parameters that are later broadcasted to vectors. In rare cases, access to single vector elements (for setting or reading one value at a time) is required. The vector data path requires arithmetical instructions (addition, subtraction, negation, multiplication, MAC), as well as comparison and maximum/minimum instructions on pairs of data vectors for sorting (e.g. for MIMO detection

Chapter 3 Scalable SIMD processor architecture

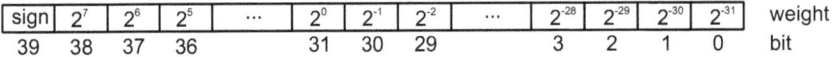

Figure 3.1: Definitions of arithmetic data types

and channel decoding). Furthermore, shift instructions for scaling and instructions for determining a shift distance (e. g. by calculating leading bits for two's complement data) are useful. Permutation instructions are required to perform a reordering of vector elements and are elaborated in more detail in section 3.1.5.

Next to these basic requirements on the instruction set, instructions and processing units that accelerate one or several applications are implemented in most SIMD-based processor architectures [vHM+05, MG08, SVPG+10, WLS+08a]. Examples for application independent specialized operations, which can be accelerated, are division, square root, and reciprocal square root. For example, a reciprocal square root is required for the QR matrix decomposition [GVL96], which is utilized in different MIMO algorithms, such as the sphere decoder in chapter 5. An example for a specialized processing unit that accelerates some algorithms is the EVP's intra vector unit (IVU, see section 2.3.1). This processing unit supports minimum and maximum search over all elements of one data vector and summation of vector elements. For example, the QRD-M MIMO detection algorithm (section 5.2) benefits from minimum search support for the detection of the most likely transmitted symbols. Although there is a potential performance gain from these specialized instructions and processing units, the focus of this paper is on the analysis and assessment of the benefit of an increased SIMD vector width and not the design of accelerators. Therefore, neither specialized instructions nor processing units have been considered during the design of the scalable SIMD processor architecture.

As an alternative to branch-based control flow, conditional instruction execution enables the conditional execution of an instruction based on the value of a Boolean condition register. If the value of the condition is *true*, the instruction is executed normally; otherwise, the instruction does not execute and the value of the destination register is left unchanged. Conditional instruction execution of SIMD operations can either be performed using scalar conditions — referred to as *predicated* execution or predication — or on an element-by-element basis using a condition vector. This case is denoted as *masked* execution or masking, as — depending on the values of the condition mask — some elements of the destination vector are updated with newly computed values, while the remaining elements are left unchanged. Predicated execution is commonly used in processor architectures that support instruction level parallelism (ILP, see section 3.1.3), because predication allows avoiding conditional control flow, which in turn limits ILP. Masked execution is useful for any kind of SIMD processor architecture as it allows more flexibility during the algorithm design by enabling the vectorization of conditional code. Furthermore, masking enables to exclude some SIMD vector elements from a computation. Hence, the SIMD vector length can be temporarily decreased.

The EVP supports masking and predication for most vector operations and predication for most scalar operations — enabling to evaluate the implemented algorithms for use cases of masking and predication. Furthermore, the EVP supports a so-called *conditional add/subtract* operation, which performs addition for mask element value *true* and subtraction otherwise. The evaluation showed that predicated execution is never used for SIMD vector operations. The only use cases are the calculation of scalar parameters and conditional pointer updates. Masked execution is used for permutation operations during the calculation of radix-2 and mixed-radix FFTs (see chapter 4), as well as for masked arithmetical operations. The conditional add/subtract operation is repeatedly used in the HSDPA spreader.

Based on the analysis, six vector and five scalar processing units have been implemented in the scalable SIMD processor architecture. The processing units and their supported operation types are

Table 3.2: Vector processing units and supported operation types

Processing unit	Abbreviation	Masking	Description of operation types
Vector arithmetic logic unit	VALU	✓	arithmetical and logic operations including shift, comparison and conditional add/subtract
Mask arithmetic logic unit	MALU		logic operations on vector masks
Vector multiply-accumulate unit	VMAC	✓	multiplication, MAC; accumulator & complex-valued data types
Vector load/store unit	VLSU		memory access, address updates
Vector permutation unit	VPU	✓	vector permutations on a vector permutation network
Vector move unit	VMU		move operations for masks / accumulator registers
Scalar exchange unit	SXU		vector element access, scalar broadcast

listed in tables 3.2 and 3.3. As predicated execution is not necessary for vector processing units, only the scalar ALU and scalar MAC support predication. The VALU, VMAC and VPU support masking for all operations; the VALU also includes the special conditional add/subtract operation. The VMAC supports both complex-valued data types and 40-bit accumulator data types for intermediate results with increased precision. Most vector and scalar operations are designed for single cycle latency as displayed in table 3.4. Exceptions are control flow operations, memory access operations, and

Table 3.3: Scalar processing units and supported operation types

Processing unit	Abbreviation	Predication	Description of operation types
Scalar arithmetic logic unit	ALU	✓	arithmetical and logical operations including shift and comparison
Predicate arithmetic logic unit	PALU		logic operations on Boolean predicates
Scalar multiply-accumulate unit	MAC	✓	multiplication, MAC
Scalar load/store unit	LSU		memory access, address updates
Branch control unit	BU		branches, zero-overhead loops

complex-valued multiplication and MAC operations. In table 3.4, the initiation interval of an operation is defined as the minimum interval between starting the execution of one operation and starting another operation on the same unit. Hence, operations with an initiation interval of one cycle can be started every clock cycle. The complex-valued multiplication and MAC operations have an initiation interval of two clock cycles, as the computation is split into two parts in two consecutive clock cycles (see equation (3.1)), but the same multipliers are used in both clock cycles to reduce the hardware overhead.

$$\operatorname{Re}\{a \cdot b\} = \operatorname{Re}\{a\} \cdot \operatorname{Re}\{b\} - \operatorname{Im}\{a\} \cdot \operatorname{Im}\{b\}$$
$$\operatorname{Im}\{a \cdot b\} = \underbrace{\operatorname{Re}\{a\} \cdot \operatorname{Im}\{b\}}_{\text{cycle 1}} + \underbrace{\operatorname{Im}\{a\} \cdot \operatorname{Re}\{b\}}_{\text{cycle 2}} \quad (3.1)$$

Branch and loop operations have an initiation interval equal to the instruction latency, because only one control flow operation can be processed at a time.

3.1.3 Instruction level parallelism

The term instruction level parallelism (ILP) is used for processor architectures that are able to issue multiple instructions per clock cycle. On a SIMD architecture, ILP support may hide the overhead for

Table 3.4: Latencies of scalar and vector instructions measured in clock cycles

Operation type	On unit	Latency	Init. interval
Load/store operations	VLSU, LSU	3	1
Complex-valued multiplication/ MAC	VMAC	2	2
Branch operation	BU	4	4
Zero-overhead loop	BU	4	4
Other operations	all units	1	1

memory access and vector permutations. Furthermore, ILP can improve the resource utilization of the SIMD processing units.

In principle, architectures that support ILP can be classified into superscalar and long instruction word (LIW) or very long instruction word (VLIW) architectures, with explicitly parallel instruction computing (EPIC) and dynamic VLIW as hybrid types [Smo02]. Superscalar architectures perform all control tasks for issuing multiple instructions in parallel in hardware. The control tasks are the grouping of independent instructions, which potentially can execute in parallel without interfering with each other, the assignment of instructions to functional units (FUs), and the actual initiation of instructions. On a LIW or VLIW architecture, all these tasks have to be performed by the programmer or — if available — compiler. The difference between LIW and superscalar architectures is illustrated in figure 3.2. LIW and VLIW architectures only differ in the number of issued parallel instructions — with no clear defined boundary between both terms; the term VLIW has been introduced by Fisher in 1983 [Fis83].

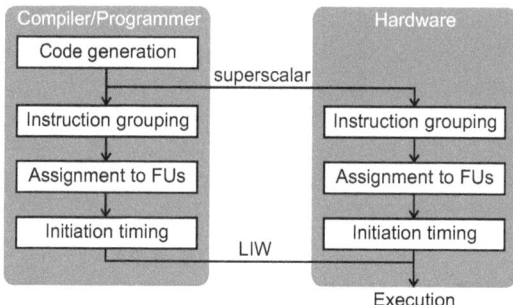

Figure 3.2: Visualization of ILP architectures based on [Smo02]. Superscalar architectures perform all control tasks for instruction parallelization in hardware, while LIW architectures require these tasks to be done by the programmer or compiler.

Due to the hardware overhead of superscalar architectures, only LIW and VLIW architectures are of interest for modern signal processors. Compared to sequential processor architectures, LIW architectures require additional register file ports to support several functional units and wider instructions that contain multiple *slots* with operations on different units. In a *fixed-length* LIW architecture (figure 3.3 on the left-hand side), slots simply occupy consecutive segments of the instruction word. Although the instruction decoding for such an architecture is of low complexity, the code size significantly grows if not all available slots can be filled with useful operations. If no useful operation can be scheduled in a slot, the slot has to be filled with a no-operation (nop). *Variable-length* LIW architectures avoid this issue by explicitly encoding the number of used slots in the instruction. The number of slots can be defined by a header or by differential encoding of slots (see figure 3.3 on the right-hand side). Differential encoding requires one additional stop bit for each slot except for the last one. The value of the stop bit defines whether another slot follows the current slot or the current slot is the last slot. The code word length can be further reduced by applying code compression techniques, for example based on Huffman codes [WC92, BNW98], Markov models [XWL02, XWL06], or lookup tables [RS03]. Due to the complexity of these code compression techniques, they have not been considered for the proposed scalable SIMD processor architecture.

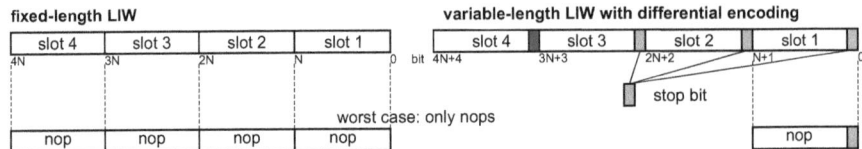

Figure 3.3: Fixed-length and variable-length LIW encoding examples for a LIW architecture with four slots

As mentioned above, LIW architectures also require an increased number of register file ports to support multiple functional units in parallel. Both area and power consumption of a register file with p ports have an asymptotic complexity of $\mathcal{O}\left(p^2\right)$ [RDK+00] (see section 3.1.4). Hence, the maximum number of LIW slots should be set to a moderate value to limit the number of ports.

In order to select an appropriate number of instruction slots, the algorithm implementations on the EVP have been analyzed. The EVP supports up to six vector and four scalar operations in one VLIW instruction. Yet, the average number of parallel operations per instruction is significantly smaller than that for all considered algorithms. The results of the analysis are depicted in table 3.5. The results show the best known implementations of these algorithms. For the radix-2 FFT, similar results have been obtained by other researchers [SM06].

The average number of parallel operations per instruction $N_{\text{par. }\varnothing}$ and the peak value $N_{\text{par. peak}}$ have been measured for the inner loops of algorithms. Furthermore, the resource utilization values of both SIMD arithmetic units (R_{VALU} and R_{VMAC}) have been calculated. Resource utilization describes the relative amount of time the unit has been active. [1] The resource utilization indicates if a speedup is possible. If any processing unit is utilized all the time, no speedup by LIW is possible without adding further processing units of the same type.

On average, all measured inner loops achieve between two and four parallel operations per instruction. Except for the HSDPA channel combiner, all inner loops achieve maximum resource utilization values close to or equal to 100 percent. The HSDPA sub frame generation kernel does not use the VALU; however, the CGU is active all the time.

Based on these results, the scalable SIMD processor architecture was designed as a LIW architecture with four parallel slots per instruction, which is more than all achieved values for $N_{\text{par. }\varnothing}$. The peak number of parallel operations has only been achieved for one or two cycles in each loop, this suggests that similar performance can be achieved by moving one (or two) operations into subsequent instructions. Hence, a slot number smaller than the peak number of parallel operations has been selected. A variable-length LIW encoding based on differential encoding has been chosen to

---

[1] The resource utilization excludes data movement operations, which do not perform useful computations.

Table 3.5: Measured ILP on the EVP for inner loops of baseband algorithms

Measured kernel	$N_{\text{par. }\varnothing}$	$N_{\text{par. peak}}$	R_{VALU}	R_{VMAC}
3 radix-2 FFT stages (with permutations)	3.654	5	92.33 %	61.54 %
3 radix-2 FFT stages (without permutations)	2.808	5	92.33 %	92.33 %
Radix-3 FFT stage	2.467	5	66.67 %	93.33 %
Radix-6 FFT stage	2.045	4	72.73 %	72.73 %
Radix-5 FFT stage	2.619	6	61.90 %	85.71 %
W-CDMA channel estimation	2.600	6	15 %	100 %
HSDPA channel combiner	2.542	5	25 %	58.33 %
HSDPA sub frame generation	3.625	5	100 % (CGU)	100 %

guarantee a small code size. The implementation is based on design examples in [CoW09b]. The number of ports for the register files will be discussed in the following section.

The instructions are encoded with a slot size of 24 bits and a variable instruction word length between 24 and 96 bits. The implemented slot encoding formats for the different operation types are displayed in figure 3.4.

3.1.4 Register files

The register files (RFs) of a SIMD architecture require a significant amount of the total energy and area. Lin et al. [LLW+06] report that 7 percent of the area of SODA are needed for the SIMD register file (16×512 bit); the SIMD register file consumes 37 percent of the total power for W-CDMA (2 Mbps) and 27 percent for 802.11a (24 Mbps). Hence, a careful dimensioning of the register files is necessary.

The area, the delay, and the power consumption of a register file depend on the number of registers N_{reg} and the number of register ports p. According to register file models based on Rixner et al. [RDK+00], the area of a register file has an asymptotic complexity of $\mathcal{O}\left(N_{reg} \cdot p^2\right)$, the delay of a register file with a large number of ports has a complexity of $\mathcal{O}\left(\sqrt{N_{reg}} \cdot p\right)$, and the power dissipation of a register file with many ports grows as $\mathcal{O}\left(N_{reg} \cdot p^2\right)$. Area, delay, and power dissipation can be reduced by using hierarchical, distributed, or streaming register files [RDK+00]. However, these optimizations are beyond the scope of this work.

The most important register file is the general-purpose SIMD register file, which is accessed by the vector arithmetic units (VALU and VMAC). In the following, design decisions for this register file are discussed. The number of read and write ports p_r and p_w depends on the number of LIW slots (see section 3.1.3) and the port requirements of functional units. p_r and p_w can be reduced by sharing ports between functional units. Yet, this technique prevents the parallel execution of these units. An analysis of algorithms on the EVP showed that vector element permutations and scalar element access or broadcast operations very seldom occur in the same clock cycle. Hence, the ports of the VPU and SXU are shared. Altogether, the port requirements of the SIMD units are as follows:

- Two read ports and one write port for the VALU

Chapter 3 Scalable SIMD processor architecture

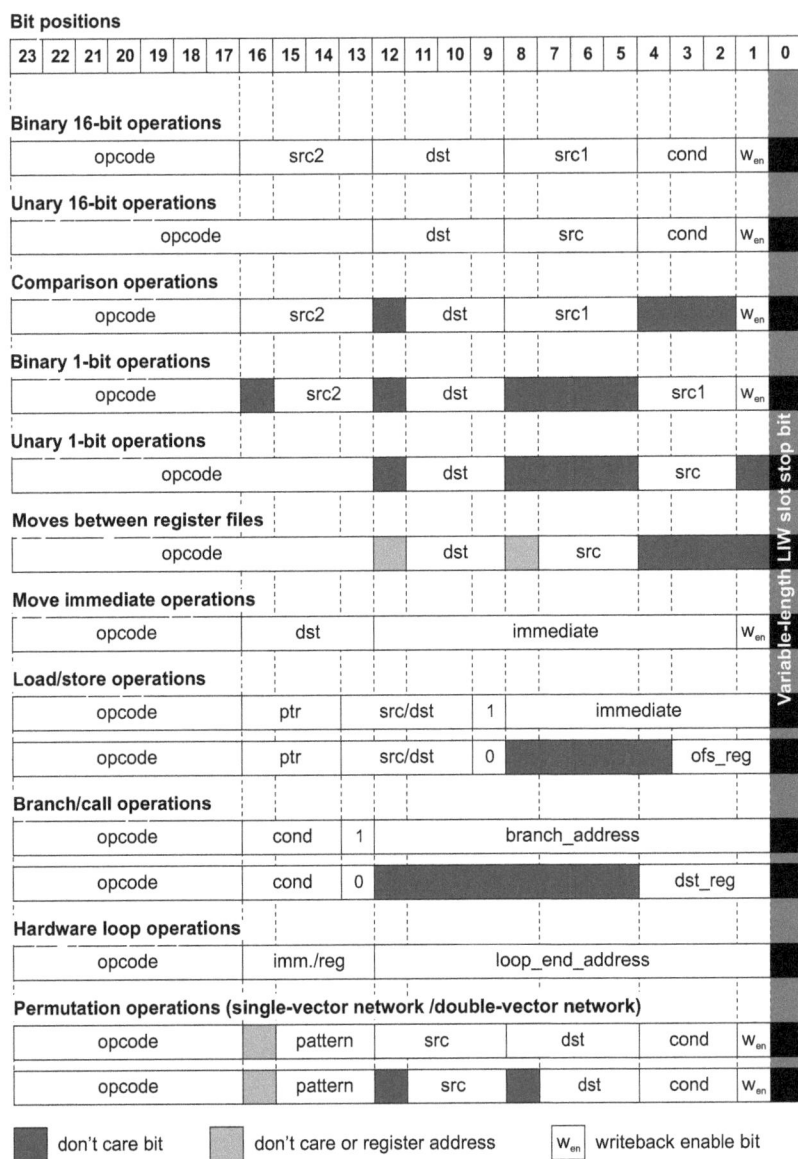

Figure 3.4: Encoding of 24-bit slots: The first bit contains the differential encoding of the instruction length, the remaining bits contain the instruction.

- Three read ports and one write port for the VMAC
- One read and one write port for the VLSU
- One read and one write port shared for VPU and SXU if a single-vector permutation network is utilized (see section 3.1.5) / two read and two write ports shared for VPU and SXU if a double-vector permutation network is utilized

Figure 3.5 shows the connections between the general-purpose SIMD register file and the SIMD units. Assuming a single-vector permutation network, which operates on one input vector and generates one output vector, seven read and four write ports are required for the register file. A double-vector permutation network supports permutations on pairs of vectors. Therefore, a double vector permutation network requires one additional read port and one additional write port, as illustrated by the dashed arrows in figure 3.5.

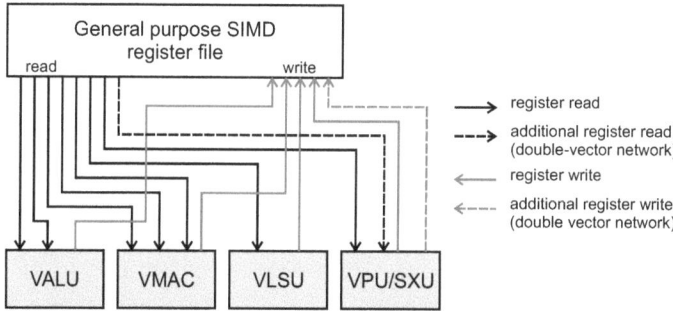

Figure 3.5: Read/write connections between the general-purpose SIMD register file and the SIMD units

The number of registers N_{reg} has been selected based on the algorithm requirements: N_{reg} should be sufficient to avoid bottlenecks for spilling values to memory, but the register file also should not be too large. As a rule of thumb, Rixner et al. [RDK+00] claim that four registers are needed per ALU and per cycle of memory latency for a LIW processor. However, the actual number of required registers depends on the processed algorithms.

The algorithm implementations on the EVP have been analyzed to identify the register demands of the algorithms. The vector register file of the EVP contains 16 registers. Except for a pair of loops in the 1024-point and 256-point FFT implementations[2], the performance of all implemented algorithms is not dominated by memory access due to a lack of registers. Hence, 16 vector registers are apparently sufficient. On the other hand, a reduction of the number of registers to eight would significantly degrade the performance as demonstrated by the exemplary discussion of radix-2 FFT loops below.

---

[2]These loops contain the processing of two consecutive radix-2 FFT stages and require memory access for loading twiddle factors on the fly.

One radix-2 butterfly operation requires two input operands (which are stored in vectors) and (at most) one twiddle factor operand. As consecutive radix-2 butterfly stages operate on different input operands, four input operands (and twiddle factor operands) need to be available for computing two consecutive radix-2 butterfly stages without spilling data to memory. Correspondingly, eight input operands (and twiddle factor operands) need to be available for grouping three radix-2 butterfly stages together (see figure 3.6). Each radix-2 stage requires one operation per vector operand on the VMAC[3] and on the VALU, while memory access always requires two operations (one load and one store operation) per vector. Table 3.6 summarizes the impact of the number of registers N_{reg} on the grouping of radix-2 butterfly stages: if only eight registers are available, the same number of operations is necessary for arithmetic operations as for loading and storing data vectors. If vectors containing twiddle factors need to be loaded from memory, the performance is dominated by memory access. For $N_{reg} = 16$, three radix-2 butterfly stages can be grouped together, improving the ratio of arithmetic to memory access operations. Hence, memory access for reading twiddle factor vectors can be hidden by arithmetic operations, which are executed in the same clock cycle during the LIW execution. An increase to 32 registers further improves the ratio of arithmetic to memory access operations, yet there is no performance gain.

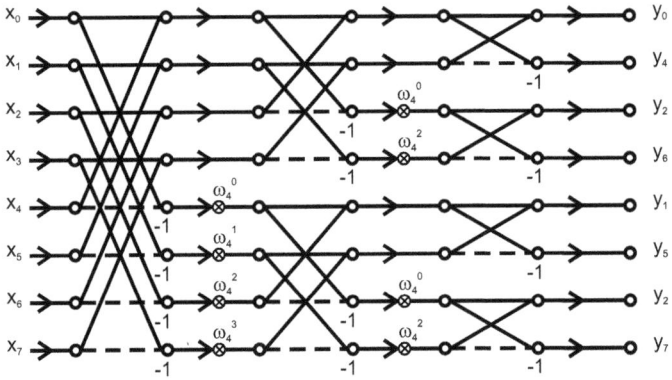

Figure 3.6: 8-point decimation in frequency (DIF) FFT

Based on the analysis above, the general-purpose SIMD register file (RF) has been implemented with 16 registers. Similar optimizations of register file sizes, as well as port sharing, have been done for the other register files. The attributes of the implemented register files are summarized in table 3.7. Most register types support bypassing to avoid read accesses and to speedup algorithms. Furthermore, the write back to the register file can be disabled for some register types on some processing units to reduce register file power. In this case, the computed result is only available via bypassing.

---

[3]Here, one complex-valued multiplication that occupies the VMAC for two clock cycles is counted as two operations.

Table 3.6: Evaluation of the grouping of radix-2 butterfly stages for different N_{reg}. S_{max} denotes the maximum number of consecutive radix-2 butterfly stages without spilling to memory; N_{vec} describes the number of vectors for the FFT operands needed to achieve S_{max}.

N_{reg}	S_{max}	N_{vec}	VALU/VMAC operations	Memory operations
8	2	4	8+8	8 + twiddle loads
16	3	8	24+24	16 + twiddle loads
32	4	16	64+64	32 + twiddle loads

Table 3.7: Register files in the scalable SIMD processor architecture. N_{bit} and N_{reg} denote the register bit-width and the number of registers in the register file, respectively. For SIMD register files, N_{bit} denotes the width of one element of the distributed register file. The number of read and write ports is described by p_r and p_w. The size of the permutation pattern registers depends on the permutation network and the SIMD width and is not listed. BYP denotes bypassing support and *dis. WB* describes optionally disabled write back.

Register file description	N_{bit}	N_{reg}	p_r	p_w	BYP	dis. WB
General-purpose SIMD RF: single-vector perm.	16	16	7	4	✓	✓
General-purpose SIMD RF: double-vector perm.	16	16	8	5	✓	✓
SIMD accumulator RF	40	2	2	2	✓	✓
Vector mask RF: single-vector perm.	1	8	5	2	✓	✓
Vector mask RF: double-vector perm.	1	8	6	2	✓	✓
Permutation pattern RF	special	8	1	1		
General-purpose scalar RF	16	16	9	5	✓	✓
Scalar predicate RF	1	8	3	1	✓	✓
Pointer RF	16	8	2	3	✓	
Range, base RFs for modulo addressing	$2 \cdot 16$	2	2	1		
Pointer offset RF	16	8	2	1	✓	

3.1.5 Permutation networks

SIMD processor architectures require permutation networks to support a reordering of vector elements. However, various types of permutation networks exist [Wak68, GL73, SS78, SMPTM79, Sie79, Par80, Pea77, Dal90, RMR+07]. These networks differ in complexity, topology, and the number of required operations to perform a desired permutation. As the permutation network should support the permutations that are required for the considered signal processing algorithms, permutations that occur frequently have been identified. These permutations should require only one permutation operation so that the permutation network does not become a performance bottleneck. Furthermore, frequently occurring permutations should be supported by special instructions that provide the necessary control signals for the permutation network. Otherwise, the performance would be limited by memory access for loading complex permutation patterns for each permutation operation.

An analysis of algorithm implementations on the EVP shows that the FFT frequently requires butterfly permutations on pairs of vectors [WBAHS08a]. The W-CDMA and HSDPA implementations [WBAHS08b, WBAHS09] also require butterfly permutations for FFTs and rotations of vector elements. Furthermore, the W-CDMA implementation requires a reversal of the ordering of vector elements, but this permutation does not occur frequently. A vector element rotation is also necessary for the decoding of quasi-cyclic LDPC codes (see section 6). Based on these results, the permutation network should directly support rotation and butterfly permutations. As the FFT requires permutations on pairs of vectors, permutation networks with different widths, i.e. one or two vectors, should be considered.

In principle, there are three relevant classes of permutation networks for SIMD processor architectures: crossbar networks, multi-stage interconnect networks (MINs), and single-stage networks. Single-stage networks only support a very limited number of permutations; more complex permutation operations require multiple consecutive permutations. Hence, this class of networks has not been considered for implementation.

Crossbar networks

Crossbar networks offer the greatest flexibility of all permutation networks; they can perform arbitrary permutations in one step. A crossbar network connects all inputs to all outputs in a two-dimensional layout (e.g. inputs in horizontal and outputs in vertical direction). The connections can be realized either by multiplexers or by using one transmission gate for each crosspoint [DOW96, DWWO96]. A transmission gate based scheme for a 4×4 crossbar is displayed in figure 3.7. Both schemes have an asymptotic complexity of $\mathcal{O}\left(N_{\text{SIMD}}^2\right)$ for an N_{SIMD}-input network.

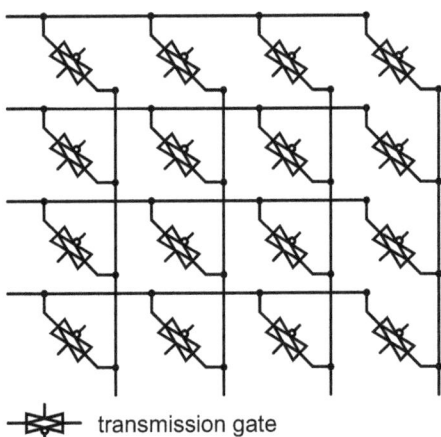

Figure 3.7: 4×4 crossbar network based on transmission gates

According to Dutta et al. [DOW96], transmission gate based crossbars should only be considered for small networks with at most 16 inputs; for larger designs, the performance is very poor. In the worst-case scenario, one input is broadcasted to all outputs; hence, one input driver has to overcome the capacitance of many outputs, which limits the performance for large network sizes. Furthermore, the area may almost grow with $\mathcal{O}\left(N_{\text{SIMD}}^3\right)$ due to the significant impact of control lines. Hence, crossbar realizations using $\log_2(N) : 1$ multiplexers for each output should be used for larger networks [DOW96]. Woh et al. [WSM+09] proposed an implementation of crossbar networks based on SRAM cells, which can reduce the power consumption of wide crossbar networks.

As crossbars directly map inputs on outputs, the control signal generation is simple. Hence, specialized butterfly and/or rotation instructions can be implemented with negligible overhead. A crossbar network can be further optimized for specific applications by removing crosspoints [RMR+07] if the full flexibility is not needed.

Multi-stage interconnect networks

MIN-based permutation networks consist of multiple consecutive permutation stages [dB87, SS78, SMPTM79, Sie79, Par80, Pea77], with each stage performing permutations on pairs of inputs. A network for N_{SIMD} inputs requires $log_2(N_{\text{SIMD}})$ stages, each consisting of $N_{\text{SIMD}}/2$ *switching elements* with two inputs and two outputs.[4] Two-function switching elements support swapping the inputs or simply forwarding inputs to outputs without a permutation. Four-function switching elements also allow broadcasting one of the inputs. The outputs of one stage of switching elements are the inputs of the next stage. Due to the multi-stage structure, the delay of MINs is proportional to the number of stages $log_2(N_{\text{SIMD}})$.

The network *topology* describes how switching elements in consecutive stages are connected to each other with different topologies referenced in literature (e.g. [SS78, Sie79]). Figure 3.8 displays three commonly used network topologies: an indirect binary n-cube network, an Omega network, and a butterfly network. In an indirect binary n-cube network [Pea77], inputs that only differ in the i-th bit are connected to the same switching element in stage i. In case of a three-stage network, this can be visualized by a cube with vertices representing inputs and edges in different dimensions representing the switching elements. An Omega network consists of multiple consecutive perfect shuffle permutation stages. A perfect shuffle operation interleaves the lower and upper half of inputs. A butterfly network performs butterfly permutations with decreasing block widths in each stage. The Omega and butterfly networks have the same topology. The Omega network in figure 3.8b can be redrawn as in figure 3.8c by exchanging the positions of switching elements F and G. Butterfly and cube network have isomorphic topologies that only differ in the ordering of stages, i.e. an indirect binary n-cube network is also an inverse butterfly network. [Par80, SS78, Sie79] show that the different networks are also functionally equivalent.

As there is only one path from each input to each output[5], MINs support a limited number of permutations. For example, the n-cube network cannot map inputs 6 and 7 on outputs 2 and 4 at the

---
[4]The term *interchange box* is also used instead of switching element.
[5]This property defines the networks as banyan networks [GL73].

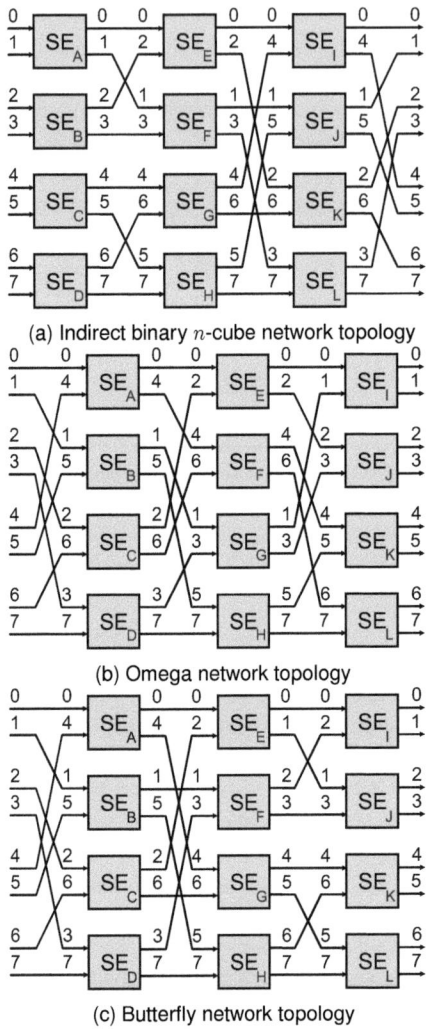

Figure 3.8: Three MIN topologies based on cube network (a), Omega network (b) and butterfly network (c) for a network with three stages and eight input values [SS78, Sie79]

same time, because both paths go through the upper output of switching element D (see figure 3.9). Arbitrary permutations require multiple consecutive permutation operations. According to Parker [Par80], any permutation of N_{SIMD} values can be realized with $\min(6, \log_2(N_{SIMD}))$ passes through an Omega network. Alternatively, a concatenation of a butterfly and an inverse butterfly network with $2\log_2(N_{SIMD}) - 1$ stages allows arbitrary permutations with $N_{SIMD} \cdot \log_2(N_{SIMD}) - N_{SIMD}/2$ switching elements. Such a network topology is denoted as a Beneš network [Ben65, Bhu09]. Waksman [Wak68] also demonstrated that the number of switching elements could be further reduced. However, Beneš networks have almost the double delay of an Omega or cube network. Hence, this type of network has not been considered.

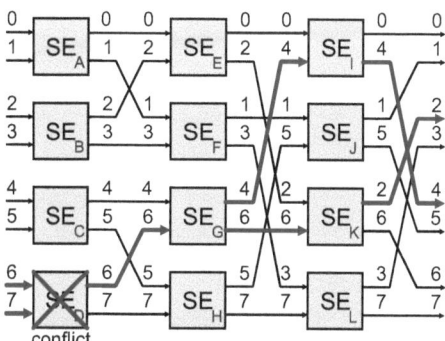

Figure 3.9: Example demonstrating the limited permutation support of MINs. Here, inputs 6 and 7 shall be mapped on outputs 2 and 4. Yet, a conflict in switching element D occurs, because only one input can be mapped on the upper output.

All MINs support butterfly permutations and rotation operations [SS78, Sie79]. Butterfly permutations simply require setting one complete stage of switching elements to swapping inputs. Hence, the control signal generation is simple. Hilewitz and Lee [HL07] developed a recursive algorithm with low complexity for determining the control bits for rotation operations on butterfly and inverse butterfly networks. Here, inverse butterfly networks have a small speed advantage, because the control signals can be calculated in parallel to the permutation stages. This is possible as the first stage of the control algorithm generates control bits for the first stage of an inverse butterfly network. The control bits for the first stage of a butterfly network however are only available after the last stage of the control algorithm.

Conclusions on permutation network support

As all presented MINs are functionally and topologically equivalent, there is no need to implement (and compare) multiple of these networks. Due to the advantages concerning the generation of control signals, an inverse butterfly network (or indirect binary n-cube network) topology is most promising as a representative of MINs. Hence, an inverse butterfly network has been implemented

Chapter 3 Scalable SIMD processor architecture

as one alternative for the permutation network in the VPU. A crossbar network offers more flexibility at a greater hardware cost. This type of network has also been implemented — enabling to compare the gain in performance and the hardware demands with the inverse butterfly network topology. As the FFT requires permutations on pairs of vectors, permutation networks with two different widths (one vector or two vectors) have been implemented. In the following, networks with a width of one vector or a width of two vectors will be denoted as *single-vector* networks and *double-vector* networks respectively. Table 3.8 summarizes the properties of the implemented networks.

Table 3.8: Summary of properties of the four implemented networks. N_{stages} denotes the number of required permutation stages, N_{SIMD} is the width of a SIMD vector in 16-bit elements.

Network description	Width	N_{stages}	Complexity
Single-vector inverse butterfly	1 vector	$\log_2(N_{SIMD})$	$\mathcal{O}(N_{SIMD}\log_2(N_{SIMD}))$
Double-vector inverse butterfly	2 vectors	$\log_2(N_{SIMD})+1$	$\mathcal{O}(N_{SIMD}\log_2(N_{SIMD}))$
Single-vector crossbar	1 vector	1	$\mathcal{O}(N_{SIMD}^2)$
Double-vector crossbar	2 vectors	1	$\mathcal{O}(N_{SIMD}^2)$

3.1.6 Overview of the SIMD processor model

Figure 3.10 shows a block diagram of the scalable SIMD processor architecture that has been realized based on the design decisions explained in the prior sections. The main features can be summarized as follow: The processor supports operations on 16-bit SIMD vector elements with additional support for complex-valued multiplications. Up to four instructions can be issued in parallel in a variable-length LIW format. The SIMD bit width is scalable from 128 bits to 1024 bits. Furthermore, four alternative permutation networks have been implemented. The resulting SIMD architecture design space consists of 16 different processor configurations and is depicted in table 3.9. Scaling is done by changing the number of 16-bit vector lanes (see figure 3.10) and adjusting the permutation network width.

Besides the SIMD architecture properties that already have been explained, the processor architecture supports some further features common to many DSP architectures. The branch control unit (BU) supports zero-overhead loops (ZOLs), i. e. special loop instructions allow to perform loops with

Table 3.9: Design space for the exploration of the scalability of SIMD processing

| Network type | Abbreviation | Supported SIMD bit widths | | | |
		128 bit	256 bit	512 bit	1024 bit
Single-vector inverse butterfly	Bfy1	✓	✓	✓	✓
Double-vector inverse butterfly	Bfy2	✓	✓	✓	✓
Single-vector crossbar	Cross1	✓	✓	✓	✓
Double-vector crossbar	Cross2	✓	✓	✓	✓

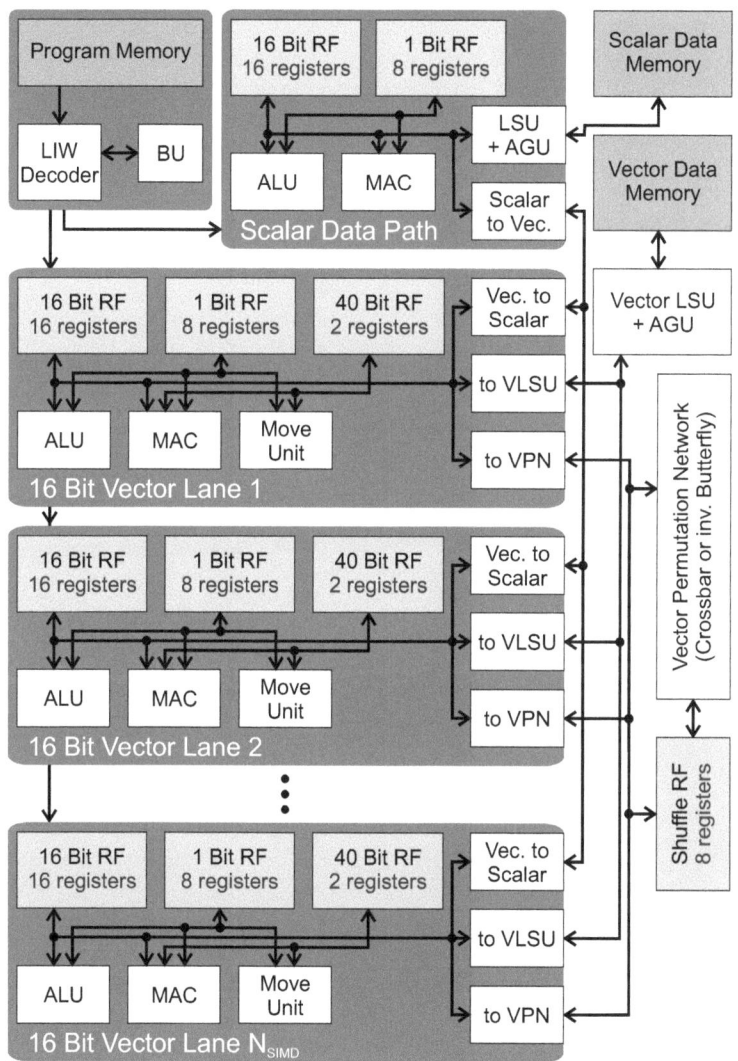

Figure 3.10: Block diagram of the scalable SIMD processor architecture

a known iteration count without the cycle overhead of testing and decrementing the loop counter and conditional branching. The processor architecture supports three nested levels of ZOLs. A ZOL is implemented by a set of three special registers that contain the loop start and end address and the loop counter. The loop counter is automatically decremented and tested in hardware.

The address generation units (AGUs), which are part of the scalar and vector load-store units (LSU and VLSU), support memory access and pointer updates in one instruction. Furthermore, modulo address operations have been implemented based on an algorithm by Prasad and Kolagotla [PK98]. Modulo addressing enables to use a contiguous address range as a circular buffer. The circular buffer is defined by a base address register and a range address register, which specifies the buffer size.

As figure 3.10 shows, the scalable SIMD processor architecture uses separate memories for scalar and vector data. The separation has been done, because both memories require different sized read and write ports (16 bit versus $N_{SIMD} \cdot 16$ bit). Two memories simplify the design and consume less power than a single big memory [LLW+06].

The SIMD processor architecture has a seven-stage pipeline. However, only operations that read data from the scalar or vector memories use the last two pipeline stages. The pipeline is displayed in figure 3.11.

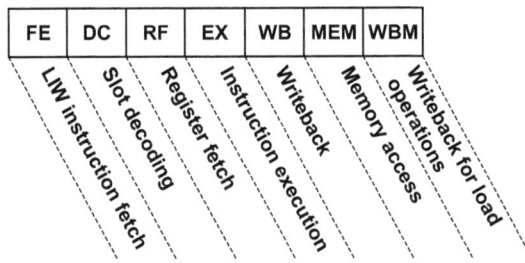

Figure 3.11: Pipeline of the scalable SIMD processor architecture

3.2 SIMD processor modeling in LISA

The scalable SIMD processor architecture has been modeled in LISA, which is an acronym of Language for Instruction Set Architecture. LISA is an architecture description language (ADL) that allows the behavioral and structural modeling of processor architectures.

Numerous ADLs have been invented, yet most only model the behavior of the processor core and do not assist the development of register transfer level (RTL) code. Examples are the ArchC language [ARB+05], developed at the University of Campinas, and Michigan Technological University's FAST [Ond05]. Target Compiler Technologies developed the retargetable IP Designer tool chain [PLGG01, GLGVP06], formerly denoted as Chess/Checkers, which supports RTL code generation from a processor model in the nML language. The Processor Designer toolkit based on LISA also supports both RTL code generation and behavioral simulation for ASIPs. Furthermore, LISA has also gained commercial acceptance [KSAF07]. Customizable processor architectures are an alternative for ASIP modeling using ADLs. Examples for customizable architectures are Tensilica's Xtensa template architecture [Ten05, Ten08] and ARC International's ARChitect Processor Configurator [ARC05]. Yet, these architectures only support customizing the instruction set in a limited way. Hence, LISA has been selected for modeling the scalable SIMD processor architecture.

The following sections explain and evaluate the modeling capabilities of LISA for SIMD processors. In section 3.2.1, the Synopsys Processor Designer toolkit, which contains development tools based on LISA, is introduced and briefly discussed. The next section describes basic modeling techniques. Section 3.2.3 explains why scalable SIMD processors cannot be directly modeled in LISA and describes an extension with a powerful macro preprocessor, which overcomes the limitations of LISA. Finally, the drawbacks of LISA as a modeling language for SIMD processors are evaluated in section 3.2.4.

3.2.1 Processor Designer toolkit overview

Processor Designer is a commercial toolkit intended for the development of ASIPs based on LISA. LISA initially has been developed as a language for fast, cycle-accurate simulation and tool generation for HW/SW co-design [ZPM96, HNBM01, Hof02, Lö4, PHZM99, Pee02]. Later on, support for the automatic generation of synthesizable RTL code has been added. Hence, the toolkit now enables to perform the complete development of a processor architecture from a single source description in LISA. The toolkit can automatically generate software development tools (assembler, linker, disassembler), fast simulation tools (cycle-accurate instruction set simulator (ISS), debugger), and RTL code in Verilog and VHDL. Furthermore, there is limited support for the automatic generation of an instruction set manual and a C compiler.

Besides requiring only one common description for the processor model, LISA has some further advantages concerning the development of processor architectures: Decoding logic for instructions does not have to be coded by the programmer; instead, the logic is automatically generated from the binary coding of instructions. This includes the decoding for LIW and VLIW architectures. Furthermore, resources like register files and pipeline-registers only need to be declared and not explicitly programmed. Hence, the architecture designer can focus on the functional behavior of processing units.

3.2.2 Processor modeling in LISA

LISA models consist of *resource* declarations and *operations* [CoW09b, CoW09a]. Resources define storage elements such as registers, memories, and pipelines. Operations describe behavior, structure, and instruction set of the processor architecture. The attributes of operations are defined in several sections:

- The DECLARE section contains local declarations, e.g. references to child operations. All operations that are referenced/used inside an operation have to be declared first.

- The CODING and SYNTAX sections define the binary coding and the assembly syntax of instructions respectively. The coding is utilized for the automatic generation of the instruction decoder; the syntax is used for assembler and disassembler.

- The BEHAVIOR and EXPRESSION sections describe the behavioral model of an operation as C code. Other operations may be executed inside a BEHAVIOR section in a mechanism similar to function inlining.

- The ACTIVATION section allows activating further operations from the current operation. Activations allow to model resource sharing and the execution of operations across pipeline stages.

Operations are organized in a hierarchy based on a chain of activations. Operations have either to be *activated* in the ACTIVATION section or *called* from the BEHAVIOR section. The difference lies in the generated code structure: Calls copy the operation into the calling operation; activations use a separate operation, which may be activated from multiple sources.

A processor can be modeled at different levels of abstraction. At the highest level of abstraction, one operation defines the complete behavior of one instruction. For RTL code generation, the model then must be refined by introducing pipeline stages, i.e. one LISA operation defines one pipeline stage of an instruction, and by sharing operations (e.g. different operations for arithmetic instructions activate one common operation that defines the behavior of an ALU). Figure 3.12 depicts an example of an operation hierarchy for a processor with a four-stage pipeline. Here, the add and sub operations share the common ALU operation; all operations share a common write back operation. Furthermore, operations and resources that belong to the same pipeline stage can be grouped together in a UNIT. Such an operation hierarchy allows to model functional units (e.g. ALU or MAC) and the sharing of resources. For example, a shared register port can be implemented by a register access in an operation that is activated from different functional units. Hence, the hardware structure of a processor can be modeled in LISA [CoW09c].

3.2.3 Extensions for modeling SIMD processors

As depicted by figure 3.10, a SIMD processor architecture contains many parallel data lanes that have the same functionality, yet operate on different data. Except for the permutation network in the VPU, the whole structure of the SIMD data path can be defined by modeling one lane and scaling the number of lanes based on the desired SIMD width.

LISA contains language elements for modeling multiple resources or operations that share a common identifier – and in case of operations a common behavior. These language elements are called *template resources* and *template operations*. Template operations and template resources are defined using angle brackets; two examples are given below:

```
1: REGISTER uint16 opnd<1 ..16>;
2: OPERATION alu<index> IN pipe.EX { .. }
```

Here, the index parameter may be used inside of the operation to index further template operations – building an operation hierarchy – or to access template resources. Template resources may only be indexed by constants or by index parameters of template operations.

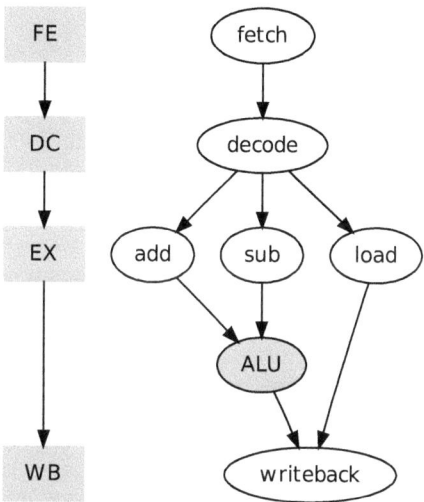

Figure 3.12: Example of an operation hierarchy with a four-stage pipeline. Grey boxes describe pipeline stages. Arrows describe the activation of operations.

Identical or similar data paths may be defined by template operations; template resources enable to define resources that are local to these data paths. These two language elements can be used to model LIW architectures with multiple processing units of the same type[6] or data paths in a SIMD architecture. However, modeling a SIMD data path is still difficult: LISA versions prior to V2009.1.0 do not support activating or calling multiple instances of the same template operation [WS09b], while newer versions at least support activations. Yet, the number of template instances that is activated has to be fixed, as each declaration and activation has to be manually implemented. Hence, a scalable architecture with a scalable number of lanes cannot be modeled directly. Additionally, permutation networks have to be programmed manually for each different network size and topology. Hence, the design of the scalable SIMD architecture in section 3.1, with four SIMD widths and four network configurations, would require maintaining 16 partially different processor models in LISA.

An approach based on the combination of LISA with GNU M4 [SPVB08, WS09b] has been chosen to overcome this issue. The M4 language is a macro processing language, which supports string manipulations, conditional evaluation, and loops. These language features allow generating a scalable number of data paths by macros that produce LISA code for template operation declarations

---

[6]This was the original purpose for introducing these language elements.

and activations (or calls). Hence, M4 is utilized as a preprocessing step. The extension of the LISA models has been realized in three steps:

- Necessary macros have been identified, implemented, and tested. The macros can be grouped into two classes: macros for the generation of permutation networks and macros for scalable SIMD data paths. Macros for scalable SIMD data paths primarily generate and control the accessing of multiple instances of template operations and resources. Macros for permutation networks generate the complete permutation network from macro calls.

- An M4 macro file containing definitions for all adjustable parameters has been generated. The parameter file defines the SIMD width, the configuration of the permutation network (topology and width), and further parameters, such as register file sizes and memory configurations.

- The M4 macros have been introduced in the LISA model. For example, a SIMD data path is added to the LISA model by implementing a template operation, which defines the behavior of the data path (e.g. OPERATION data_path<index> { ...}), and instantiating and activating the scalable data path using the SIMD_INSTANCE(data_path) and SIMD_ACTIVATION(data_path) macros in the parent operation.

The preprocessing and compilation of the LISA model have been integrated into makefiles that enable to alter the M4 parameter configuration by adding parameters to *make*.

Preprocessing using M4 macros has not only been done for SIMD support — further macros have been implemented for simplifying the LISA model structure and for circumventing bugs in LISA. For example, the LISA code for operand bypassing and saturation logic[7] is generated from macros. The binary opcode encoding of instructions for one processing unit (e. g. 0b1001001 for a scalar addition) is generated by defining a base opcode and using opcode offset or increment macros. Furthermore, macros enable generating different LISA code for simulation and RTL code generation for features of LISA that are not supported for both.

The effort for developing and maintaining a scalable SIMD model in LISA with and without M4 macro support cannot be measured directly. Yet, the number of source lines of code (SLOC) can be used as an indicator for the model complexity. Table 3.10 shows measured SLOC excluding comments and empty lines. The number of lines of code has been measured before and after macro expansion. The reference value includes the macro files. In the best-case scenario (128-bit SIMD processor and single-vector crossbar network), the code size after macro expansion is only approximately 67 percent higher than before macro expansion. However, in the worst-case scenario (1024 bit SIMD bit width and double-vector butterfly network), the model requires almost 15 times as many SLOC after macro expansion.

3.2.4 Drawbacks of LISA as a modeling language for SIMD processors

While LISA enables the rapid development of ASIPs, some issues occur when more complex SIMD processors are modeled. The first limitation, the lack of mechanisms that support scalable data

---

[7]LISA directly supports saturation logic, however the generated VHDL code may contain errors.

Table 3.10: Source lines of code measured for LISA model before and after M4 macro expansion

Measurement	SIMD width	Network	SLOC	Normal. SLOC
Before macro expansion	all	all	7292	1.00
After macro expansion: best case	128 bit	cross1	12179	1.67
After macro expansion: worst case	1024 bit	bfy2	108986	14.95

paths, can be overcome by the proposed extension with M4 macro processing. The remaining major issues are bugs that result in erroneous simulator behavior or RTL code and the code quality of the generated RTL code.[8]

Originally, LISA has been developed as a language for fast and accurate instruction set simulation (ISS). Support for RTL code generation has been added later. Hence, the simulation capabilities are more mature than RTL synthesis capabilities. The most important shortcomings of the RTL model are listed below:

- The LISA processor generation tools apparently do not recognize that template operations share the same behavior. Therefore, all instances of template operations are compiled separately. This results in increased runtime and code size for wider SIMD models.

- The modeling capabilities for hierarchical RTL models, where resources and logic are grouped locally, are insufficient. While LISA supports assigning operations and resources to functional units using the UNIT resource, bugs limit the usefulness of this mechanism: Register arrays and pipeline registers cannot be assigned to units, while operations need to belong to the same pipeline stage. All resources and operations that are not assigned to user-defined units are grouped in default units for pipeline stages. Hence, it is impossible to model a desired RTL hierarchy in LISA. The impact on gate level synthesis is discussed in section 3.4.3.

- Processor Designer can automatically generate synthesis scripts for commonly used synthesis tools, such as Synopsys Design Compiler. Yet, the generated scripts use outdated commands and are not suitable for hardware synthesis.

Despite the above-mentioned shortcomings of the Processor Designer tools, LISA still is an adequate language for ASIP design and design space exploration (DSE), because it facilitates a fast development of processor architecture and tools. Furthermore, new instructions can be added and tested with little programming effort. Many of the listed shortcomings only occur for complex models that use template resources and operations extensively — like the proposed scalable SIMD processor architecture. The modeling of such processor architectures obviously is not the design focus of the LISA tool set.

---

[8]More information about the processor development with LISA can be found on the Synopsys processor development website (http://www.synopsys.com/Tools/SLD/ProcessorDev).

3.3 Vertical-horizontal vector processing as an alternative for LIW

LIW and especially VLIW architectures have some disadvantages compared to an architecture that issues one instruction per clock cycle: Firstly, the instruction decoder is more complex. Secondly, more read and write ports are needed for the register files, which leads to increased area and power demands and an increased delay [RDK+00].

An alternative for LIW and VLIW processing based on *vertical-horizontal vector* processing is sketched in the remainder of this section. Vertical-horizontal vector processing can potentially overcome the above-mentioned shortcomings of LIW processing at the cost of reduced flexibility.

The term vertical-horizontal vector processing — or more precise *vertical-horizontal processing pipeline vector computer* — has been defined by Gao et al. [GZYC86]. The term describes parallel and time-sequential vector processing in an analogy to two-dimensional spacial processing. The horizontal component describes the parallel processing of data vectors, i. e. multiple processing units process data in parallel. The vertical component refers to the iterative processing of data vectors over time, in one instruction. In consequence, vertical-horizontal processing defines a technique, where data vectors that are too long to be processed in parallel are segmented into blocks that fit into the parallel data path. The segments are then processed sequentially. The underlying concept has been applied to early vector supercomputers, such as the Cray-1 [Rus78]. Figure 3.13 illustrates the idea. In the first clock cycle, an operation on the vector MAC unit is started, which runs for multiple clock cycles. In the next clock cycle, an operation on the vector ALU is initiated. Both operations run in parallel for the next clock cycles. On a LIW processor architecture, new vmul and vadd instructions would have to be issued in each cycle to achieve the same behavior.

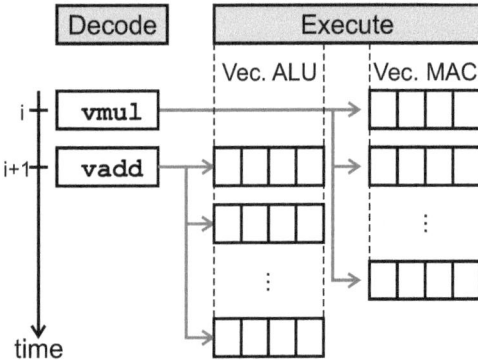

Figure 3.13: Example for vertical-horizontal vector processing with two vector units

The concept of a vertical-horizontal vector processing architecture for SDR has been examined in a diploma thesis [Lec09, in German] and is explained below in section 3.3.1. Some performance benchmarking based on three SDR algorithms is described in section 3.3.2.

3.3.1 Vertical-horizontal vector processing for SDR

The basis for the vertical-horizontal SDR architecture is the scalable SIMD processor architecture developed in section 3.1. Modifications have been done on the instruction fetch/decode mechanism and the organization of register files.

ILP is no longer achieved by issuing multiple instructions per cycle in a long instruction word; instead, one instruction that may iterate for multiple clock cycles is issued in each cycle. As operations that start successively overlap, parallelism is achieved. The iteration count of an instruction is explicitly encoded in the instruction word. As an operation iterates over multiple clock cycles, multiple source and destination registers are required — assuming that the size of one register remains the same. This can be achieved by simply incrementing the register address, yet, in this case, the number of required register file ports does not decrease compared to a LIW architecture. Instead, the proposed register file organization is based on the assumption that successive instruction iterations usually do not need to access the same data values. Hence, successive iterations may be mapped on different register file banks. A small number of register banks is provided; instructions iterate through these register banks in a cyclic manner (see figure 3.14). In each clock cycle, each register bank is accessed by a single functional unit, which reduces the number of required ports. If data from one register bank is needed for the calculations in a different bank, it has to be explicitly transferred in an instruction.

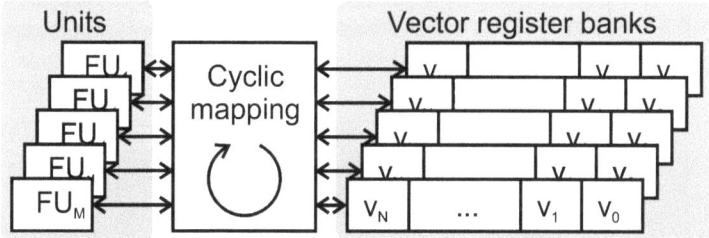

Figure 3.14: Cyclic mapping of FUs on register banks: In each cycle, the read and write ports are mapped to a single units. The FUs iterate through register banks in a cyclic manner.

The partitioning of the register file leads to reduced area and power demands. This effect is demonstrated by the following comparison of a four-way LIW architecture and a vertical-horizontal vector processing architecture with four register banks and, hence, up to four parallel instructions. The asymptotic area and power complexity for a monolithic register file (LIW case) is $\mathcal{O}\left(N_{reg} \cdot p^2\right)$ [RDK+00], with p denoting the number of ports. The asymptotic complexity of a partitioned register file (vertical-horizontal vector processing case) is $\mathcal{O}\left(N_{reg} \cdot N_b \cdot p_b^2\right)$, with N_b and p_b denoting the number of banks and the number of ports per bank respectively. N_{reg} describes the number of registers per bank. Table 3.11 shows a comparison of different register file configurations. Normalized energy and area are computed based on the model of the asymptotic complexity, normalization is done based on a monolithic register file for a four-way LIW architecture with 16 registers and 12 ports (four FUs, two read ports and one write port per unit). As energy and area depend on the squared number of ports, the reduced number of ports for a partitioned register file has a significant influence. The table also shows a combination of LIW and vertical-horizontal vector processing with two-way LIW and two register banks (eight registers and six ports per register bank). This architecture configuration requires more area and power than the pure vertical-horizontal vector processing configurations, yet in comparison to a four-way LIW architecture with monolithic register file, the area and power demands are still reduced by 75 percent.

Table 3.11: Model-based register file comparison of monolithic and partitioned register files

Description	N_{reg}	N_b	p/p_b	Normalized area & power
Monolithic register file, four-way LIW	16	1	12	1.000
Partitioned register file	4	4	3	0.0625
	8	4	3	0.125
Partitioned register file, two-way LIW	8	2	6	0.250

3.3.2 SDR algorithm performance

While a vertical-horizontal vector processing architecture with a partitioned register requires less power and area for the register file than a LIW architecture, the processor architecture is also less flexible as only one instruction can be started in each clock cycle. The effects of the reduced flexibility have been studied for three different signal processing kernels [Lec09]: matrix-vector product, a 16-point FFT, and Viterbi decoding on a 64-bit SIMD processor. The performance of an architecture without ILP support, a four-way LIW architecture, and a vertical-horizontal vector processing architecture with four register banks have been measured. The results are summarized in table 3.12. The performance of the vertical-horizontal architecture is only slightly worse than the LIW architecture performance. Furthermore, the number of instructions can be significantly reduced.

Table 3.12: Performance comparison of LIW and vertical-horizontal vector processing architectures: All figures are normalized to the results of an architecture without ILP support.

Kernel	Four-way LIW		Vertical-horizontal	
	Speedup	Instructions	Speedup	Instructions
Matrix-vector product	1.143	1.00	1.143	0.33
16-point FFT	1.542	1.00	1.423	0.44
Viterbi Trellis computation	1.714	1.00	1.600	0.56

Vertical-horizontal vector processing may offer performance similar to a LIW architecture with reduced area and power demands for the register files and the decoding of instructions. However, the performance of more complex applications could be much worse, because there is only a benefit of vertical-horizontal vector processing if the same operation can be performed multiple times on different data. If different operations have to be performed each cycle, the performance degrades to the performance of an architecture without ILP. This effect can partially be compensated by combining LIW execution and vertical-horizontal vector processing. Yet, further studies are needed to analyze performance and area and power demands for this case. Furthermore, other, more complex applications should be analyzed. However, these studies are beyond the scope of this work.

3.4 SIMD architecture analysis methodology

The analysis and evaluation of the SDR algorithms in chapter 7 is based on performance and energy consumption of a synthesized gate level model using a standard cell library. In the following, the used methodology and its limitations are shown. A block diagram of the analysis methodology is depicted in figure 3.15.

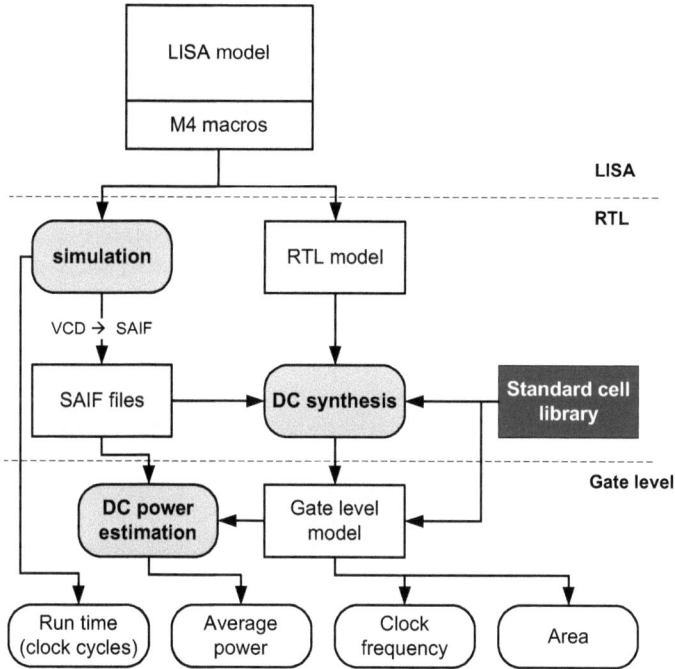

Figure 3.15: Block diagram showing the synthesis and analysis methodology

3.4.1 Processor model synthesis

Power, area, and energy consumption figures are extracted from a gate level model based on a 90 nm standard cell library by Taiwan Semiconductor Manufacturing Company (TSMC) [Tai06]. The synthesis is done using Synopsys Design Compiler (DC) Ultra [Syn09a, Syn09b]. The input for the synthesis is an RTL model generated from the LISA model and the switching activity from selected SDR algorithms. The algorithms are listed in table 3.13. As the cells in the standard cell library are characterized for internal power, this approach enables to optimize for power consumption during the synthesis process.

Table 3.13: SDR algorithms used for power optimization during synthesis

Algorithm	Description
FFT 1024	1024-point radix-2 FFT
FFT 256	256-point radix-2 FFT
FSD 16	Fixed-complexity sphere decoder for 16-QAM and 4×4 MIMO
LDPC 32, 5/6	LDPC decoder with z factor 32 and code rate 5/6
LDPC 64, 5/6	LDPC decoder with z factor 64 and code rate 5/6

The synthesis strategy is based on a *top-down* approach.[9] Top-down synthesis automatically takes care of dependences between sub-designs and enables to ungroup the design hierarchy generated from LISA. This is necessary, as a desired design hierarchy cannot be efficiently model in LISA (see section 3.2.4). The major drawback of a top-down synthesis is the significantly increased memory requirement, as the complete model has to be kept in memory during the whole synthesis process.
The synthesis has been done based on a timing analysis for both worst and best case corners (see table 3.14) with a target clock frequency of 300 MHz. The clock frequency was selected to allow comparing algorithm performance on the proposed scalable SIMD processor architecture and on the EVP, which is designed for a working frequency of 300 MHz. The design is optimized by applying clock gating and operand isolation techniques.

Table 3.14: Standard cell library operating conditions

Corner	Voltage	Temperature	PMOS/NMOS process
Best case	1.32 V	0°C	fast/fast
Nominal case	1.2 V	25°C	typical/typical
Worst case	1.08 V	125°C	slow/slow

Synthesis is done using *topographical technology*. Topographical technology performs a virtual layout using both cell library and physical library data. The topographical mode uses the same placement and optimization technologies used in Synopsys place and route tools. This approach allows an accurate prediction of post-layout timing, area, and power [Syn09a]. According to [Syn07], power results from Design Compiler in topographical mode are usually within 10 percent of the final layout results. Traditional synthesis approaches use simple *wire load* models to describe interconnect wiring

---

[9]In a top-down approach the whole processor model is compiled together. In a bottom-up approach sub-designs at the bottom of the design hierarchy are compiled first with the synthesis process iterating through the hierarchy levels until a stable solution is found.

properties, do not synthesize high fan-out[10] nets, and do not model clock trees [Syn09b]. The virtual layout in topographical layout allows avoiding wire load models and automatically synthesizing high fan-out nets. Furthermore, the clock tree power is estimated.

3.4.2 Extraction of area, power, energy and performance figures

Area and timing results can be directly reported from the synthesized gate level processor model. The area report contains the total cell area in μm^2 excluding the area required for interconnections. Timing analysis reports the path delay of critical paths, which allows checking whether the desired target frequency can be achieved.

If the frequency constraint is fulfilled, the runtime of applications can be calculated based on cycle count simulations in LISA. For the implemented SDR algorithms, the cycle count can also be directly obtained from the assembly code, because none of the algorithms exhibits data dependent control flow and the instruction execution on the processor is deterministic.

Power figures can be obtained from average power analysis in Design Compiler. The power analysis requires the gate level model and the switching activity of the algorithms. The switching activity can be measured during gate level simulations or based on the RTL switching activity, which has also been used for power optimization during synthesis. RTL simulation of switching activity requires a name mapping from RTL to gate level signals, which may lead to inaccuracies due to mapping conflicts, but is significantly faster. As an accurate power estimation also requires huge amounts of input data, which further increases the runtime of simulations, the latter approach based on RTL switching activity has been selected for the power analysis. The power analysis reports dynamic and static power consumption (see [KFA+07]). Dynamic power contains the switching power for charging and discharging capacitive loads and the internal power due to short-circuit or crowbar current, which flows when both the PMOS and NMOS transistor in a cell are open. Figure 3.16 visualizes these effects. Static power describes power consumed by leakage currents [KFA+07].

Energy consumption can be calculated from the average power and the algorithm runtime. In chapter 7, comparisons between differently configured SIMD processors (i. e. different SIMD widths or interconnect networks) are done using normalized energy, area, and performance. Normalization is done based on the results for a 128-bit SIMD processor with a single-vector butterfly network, which has the lowest complexity of all considered SIMD processors.

3.4.3 Limitations of the proposed methodology

The used analysis methodology has some drawbacks that need to be discussed. The limitations concern the standard cell approach, the accuracy of area and power results, and the ungrouping of sub-designs.

A standard cell library may not be the most efficient approach for the design of permutation networks: A full-custom design based on transmission gates should be more power and area efficient for small

---

[10]Fan-out describes the ability of a logic gate to drive a number of inputs.

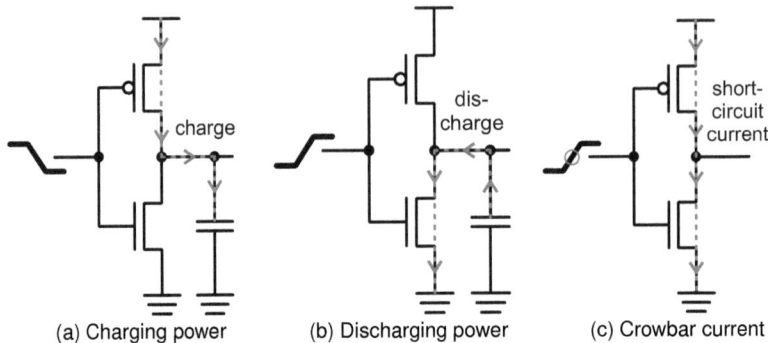

Figure 3.16: Dynamic power consumption: The dynamic power consumption results from switching power (a), (b) and internal power (c)

crossbar sizes [DOW96], while custom multiplexer-based designs should be better for inverse butterfly and large crossbar networks. However, the focus of this paper is on the comparison of different SIMD processor architectures. The additional effort for a full-custom design of multiple different permutation networks is not justified, as it would only enable slightly more accurate power and area estimates. Considering that all algorithms in chapters 4, 5 and 6 are optimized for a minimal number of permutation operations, the overall impact on the energy consumption should be low.

The second drawback of the proposed methodology is the limited accuracy of area and power estimates, because area and power are estimated at the gate level and not after place and route. Place and route have been omitted, because neither the required tools nor the required libraries have been available. Yet, the gap between power and area estimates at the gate level and results after final placement can be reduced by using the topographical synthesis mode, which produces a virtual layout. According to [Syn07], the errors in power consumption should be less than 10 percent after synthesis in topographical mode. Furthermore, a relative comparison of SIMD processors with different SIMD widths should be even more accurate, as all processors use the same data paths and the synthesized gate level processor models should therefore be similar — leading to similar power estimation errors.

The area and power estimates are also inaccurate, because the scalar and vector data memories and the program memory have not been synthesized. However, this is a conventional approach, because memories are usually designed using special memory generators and not by synthesizing standard cells. Estimated power and area figures for memories based on the on-line tool CACTI [TMAJ08] are presented in chapter 7.1.3.

The third limitation of the selected approach occurs due to the RTL design hierarchy generated from LISA. As LISA unit definitions do not work properly for all resources and operations, the generated design hierarchy needs to be ungrouped during the synthesis process. This does not necessarily affect the quality of the generated gate level code, yet estimating power and area for a specific functional unit or register file may be difficult, because the corresponding sub-design might have been ungrouped; hence, the gates cannot be attributed to a specific functional unit anymore. This behavior is a major drawback of a modeling approach based on LISA, which cannot be overcome. Yet, there is no impact on comparisons between different SIMD processors.

Chapter 4

Radix-2 and mixed-radix FFTs for OFDM-A and SC-FDMA

The FFT is the most important processing step of block modulation schemes based on orthogonal frequency division multiplexing (OFDM). Section 4.1 contains a brief overview of OFDM-based systems for multiple users and motivates the need for radix-2 and mixed-radix FFT algorithms. Afterwards, radix-2 and mixed-radix FFT algorithms for SIMD processors are developed and evaluated. In section 4.2, the matrix representation of the FFT is explained and vectorizable formulas are introduced. Next, an overview of related work on vector FFT algorithms is given in section 4.3. Section 4.4 comprises the mathematical derivation and discussion of the radix-2 and mixed-radix FFT algorithms. The radix-2 FFT algorithm enables an efficient vectorization if the length of the FFT is at least twice the vector length. The mixed-radix FFT algorithm requires the length of the FFT to be a multiple of the squared vector length. Both algorithms minimize the number and complexity of vector element permutation stages. In section 4.5, the implementation of the FFT algorithms on the scalable SIMD processor architecture is explained. The next section (section 4.6), comprises an analysis of the performance of the implemented FFTs, focusing on the scalability, and a comparison to other FFT implementations for OFDM systems. Conclusions are drawn in section 4.7.

4.1 OFDM-A and SC-FDMA

OFDM is a block modulation scheme that defines a signal in frequency domain. At the transmitter side, the signal is transformed into time domain using an IDFT for each OFDM symbol; the receiver performs a DFT to recover the symbol. Each OFDM symbol consists of many sub-carriers, which overlap in frequency domain, but are orthogonal to each other in time domain. Each sub-carrier represents an independent narrowband flat fading channel; hence, OFDM is not prone to frequency-selective fading. For an OFDM symbol size M, each sub-carrier operates at $1/M$ times the bit rate of the OFDM symbol, which leads to robustness in multipath environments [MLG06]. OFDM symbols are usually preceded by a cyclic prefix (CP), which prevents intersymbol interference (ISI) [SBM+04]. Due to the orthogonality of sub-carriers, the complexity of channel estimation techniques or MIMO detection techniques for OFDM systems is significantly lower than, for example, the complexity for CDMA systems. The main drawback of the OFDM technique is the high peak-to-average power ratio (PAPR) [MLG06], which results from the fact that the signal is defined in frequency domain, yet

Chapter 4 Radix-2 and mixed-radix FFTs for OFDM-A and SC-FDMA

transmitted in time domain, leading to fluctuations in the transmitted waveforms. Due to the high PAPR, amplifiers with a high dynamic range are required, which reduces their power efficiency. Orthogonal frequency division multiple access (OFDM-A) is a multi-user scheme based on the OFDM technique — each user is assigned a different subset of sub-carriers. The orthogonality of sub-carriers enables multiple users to transmit or receive independent data streams in parallel. Each user operates at a reduced data rate compared with the maximum data rate for single user transmission.

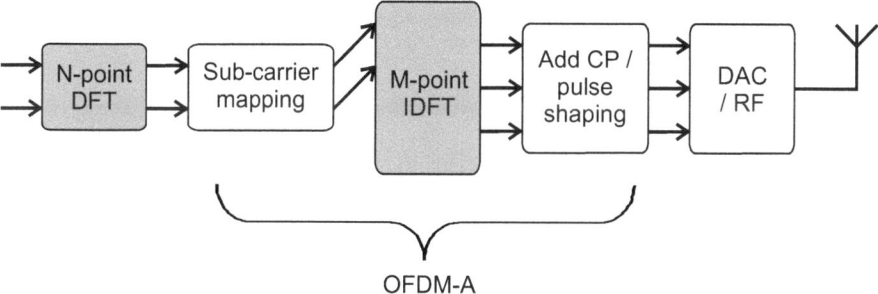

Figure 4.1: Block diagram of an SC-FDMA transmitter

Single carrier frequency division multiple access (SC-FDMA) is a multi-user scheme based on OFDM-A with an improved PAPR. Figure 4.1 shows a block diagram of an SC-FDMA transmitter. An OFDM system consists of the M-point IDFT and the adding of the CP; in an OFDM-A system, sub-carriers also have to be assigned to different users. SC-FDMA extends OFDM-A by a preprocessing step with a short N-point DFT ($N < M$); hence, SC-FDMA is also denoted as DFT-spread OFDM-A. The signals in SC-FDMA are defined in time domain and later also transmitted in time domain, which limits fluctuations in the transmitted waveform and reduces the PAPR [MLG06]. Therefore, SC-FDMA is utilized in the uplink of LTE systems [Tec09b], enabling to reduce the power consumption of mobile terminals. The LTE downlink uses OFDM-A.

LTE FFT sizes

The FFT sizes and the corresponding bandwidths for OFDM-A in the LTE downlink and the OFDM-A part of the uplink (i. e. the IDFT at the transmitter and the DFT at the receiver) are listed in table 4.1. The data carriers are assigned to the users. Information is transmitted in 10 ms frames, which consist of 20 slots. Each 0.5 ms slot contains six OFDM symbols in long CP mode and seven OFDM symbols in short CP mode.

The maximum DFT size for the N-point DFT in SC-FDMA is given by the number of data carriers in table 4.1. Each user is assigned a number of resource blocks N_{RB}; each resource blocks consists of 12 data carriers. The DFT size is further restricted to powers of two, three, and five to limit the complexity of the DFT [TSG06]:

$$N = 12 \cdot N_{RB} = 2^a \cdot 3^b \cdot 5^c \qquad (4.1)$$

Table 4.1: IDFT sizes at the transmitter side in LTE [Tec10]

Bandwidth [MHz]	1.4	3	5	10	15	20
IDFT size	128	256	512	1024	1536	2048
Data carriers	72	180	300	600	900	1200

Thus, LTE requires algorithms for radix-2 and mixed-radix FFTs.

4.2 Matrix representation of the FFT

Equation (4.2) defines an N-point DFT as a sum of the complex-weighted inputs.

$$y_k = \sum_{j=0}^{N-1} x_j \cdot e^{-2\pi i \cdot \frac{jk}{N}} \quad \text{for } k = 0, \ldots, N-1 \qquad (4.2)$$

The inverse DFT requires different weights ($e^{+2\pi i \cdot \frac{jk}{N}}$) and a scaling factor, but is in principle identical to the DFT. Hence, the further discussion focuses solely on the DFT. FFT algorithms can be derived from equation (4.2) by decomposing the sum into smaller sums and by performing index manipulations (e.g. [CT65, Sin67, Rad68]). Yet, this representation is only appropriate for a DFT decomposition into few factors.

The matrix form of the DFT, defined by equation (4.3), is better suited for arbitrary FFT decompositions and enables a compact notation for FFT algorithms.

$$\mathbf{y} = \mathbf{W}_N \cdot \mathbf{x} \qquad (4.3)$$

The elements of the $N \times N$ DFT matrix $\mathbf{W}_N$ are the complex roots of unity:

$$[\mathbf{W}_N](j,k) = \omega_N^{jk} = e^{-2\pi i \cdot \frac{jk}{N}} \quad \text{for } j,k = 0, \ldots, N-1 \qquad (4.4)$$

FFT algorithms in matrix form exploit symmetries in the DFT matrix that enable decomposing the matrix into a product of multiple sparse matrices.

4.2.1 Basic DFT decomposition for two factors

The basic decomposition of a DFT matrix is defined by equation (4.5); the notation is based on Temperton [Tem83].

$$\mathbf{W}_{pq} = (\mathbf{W}_q \otimes \mathbf{I}_p) \cdot \mathbf{P}_q^p \cdot \mathbf{D}_q^p \cdot (\mathbf{W}_p \otimes \mathbf{I}_q) \qquad (4.5)$$

In the following, the elements and operators used in equation (4.5) are explained. The matrices $\mathbf{I}_p$ and $\mathbf{I}_q$ define identity matrices of order p and q respectively.

Chapter 4 Radix-2 and mixed-radix FFTs for OFDM-A and SC-FDMA

$\mathbf{P}_q^p$ defines a $pq \times pq$ permutation matrix, which contains exactly one non-zero element (value one) in each row and each column. The elements of the permutation matrix are defined by:

$$\left[\mathbf{P}_q^p\right](j,k) = \begin{cases} 1 & \text{for } j = r \cdot p + s,\, k = s \cdot q + r \\ 0 & \text{otherwise} \end{cases} \quad (0 \leq r < q,\, 0 \leq s < p) \tag{4.6}$$

A multiplication with permutation matrix $\mathbf{P}_q^p$ results in a stride by q reordering of elements [GCT92], i.e. the elements of input vector $\mathbf{x} = (x_0, x_1, \ldots, x_{pq-1})^T$ are reordered as $\mathbf{y} = (x_0, x_q, x_{2q}, \ldots, x_{pq-q}, x_1, x_{q+1}, \ldots, x_{pq-1})^T$.

$\mathbf{D}_q^p$ defines a $pq \times pq$ diagonal matrix containing the roots of unity or twiddle factors.

$$\left[\mathbf{D}_q^p\right](j,k) = \begin{cases} \omega_{pq}^{s \cdot m} & \text{for } j = k = s \cdot q + m \\ 0 & \text{otherwise} \end{cases} \quad (0 \leq m < q,\, 0 \leq s < p) \tag{4.7}$$

The operator $\otimes$ defines a Kronecker matrix product (short Kronecker product) by:

$$\mathbf{A} \otimes \mathbf{B} = (a_{ij} \cdot \mathbf{B}) \tag{4.8}$$

For square matrices $\mathbf{A}$ and $\mathbf{B}$ of order p and q, respectively, the elements of the output matrix are defined by:

$$[\mathbf{A} \otimes \mathbf{B}](jq + l, kq + m) = [\mathbf{A}](j,k) \cdot [\mathbf{B}](l,m) \tag{4.9}$$

For example, a Kronecker product of an arbitrary 2×2 matrix $\mathbf{A}_2$ with a 3×3 identity matrix can be written as:

$$\mathbf{A}_2 \otimes \mathbf{I}_3 = \begin{bmatrix} A_{1,1} & & & A_{1,2} & & \\ & A_{1,1} & & & A_{1,2} & \\ & & A_{1,1} & & & A_{1,2} \\ A_{2,1} & & & A_{2,2} & & \\ & A_{2,1} & & & A_{2,2} & \\ & & A_{2,1} & & & A_{2,2} \end{bmatrix} \tag{4.10}$$

4.2.2 Formula manipulation rules for the DFT in matrix form

FFT algorithms for $\mathbf{W}_N$ can be developed by repeatedly applying equation (4.5), decomposing the DFT matrix until the sizes of all smaller DFT matrices $\mathbf{W}_p$, $\mathbf{W}_q$ are prime numbers, and by applying formula manipulations that reorder or transform matrices. Below, important formula manipulation rules are listed [Tem83, FP03, FVP07]. The matrix $\mathbf{A}$ is assumed a $p \times p$ matrix and the matrices $\mathbf{B}$ and $\mathbf{C}$ are assumed $q \times q$ matrices.

$$\mathbf{I}_p \otimes \mathbf{I}_q = \mathbf{I}_{pq} \tag{4.11}$$

$$(\mathbf{A} \otimes \mathbf{B}) = (\mathbf{A} \otimes \mathbf{I}_q) \cdot (\mathbf{I}_p \otimes \mathbf{B}) \tag{4.12}$$

$$(\mathbf{A} \otimes \mathbf{B}) = (\mathbf{I}_p \otimes \mathbf{B}) \cdot (\mathbf{A} \otimes \mathbf{I}_q) \tag{4.13}$$

$$(\mathbf{B} \otimes \mathbf{I}_p) \cdot (\mathbf{C} \otimes \mathbf{I}_p) = (\mathbf{B} \cdot \mathbf{C}) \otimes \mathbf{I}_p \tag{4.14}$$

$$\mathbf{A} \otimes (\mathbf{B} \cdot \mathbf{C}) = (\mathbf{A} \otimes \mathbf{B}) \cdot (\mathbf{A} \otimes \mathbf{C}) \tag{4.15}$$

$$\mathbf{P}_q^p \cdot (\mathbf{A} \otimes \mathbf{B}) = (\mathbf{B} \otimes \mathbf{A}) \cdot \mathbf{P}_q^p \tag{4.16}$$

$$\mathbf{P}_q^{pr} = \left(\mathbf{P}_q^p \otimes \mathbf{I}_r\right) \cdot \left(\mathbf{I}_p \otimes \mathbf{P}_q^r\right) \tag{4.17}$$

$$\mathbf{P}_{qr}^p = (\mathbf{I}_q \otimes \mathbf{P}_r^p) \cdot \left(\mathbf{P}_q^p \otimes \mathbf{I}_r\right) \tag{4.18}$$

$$\mathbf{P}_q^p \cdot \mathbf{P}_p^q = \mathbf{I}_{pq} \tag{4.19}$$

$$(\mathbf{A} \otimes \mathbf{B})^{-1} = \mathbf{A}^{-1} \otimes \mathbf{B}^{-1} \tag{4.20}$$

$$\mathbf{P}_q^1 = \mathbf{P}_1^q = \mathbf{I}_q \tag{4.21}$$

$$\mathbf{D}_q^1 = \mathbf{D}_1^q = \mathbf{I}_q \tag{4.22}$$

The following equation defines an abbreviation for a similarity transformation by an arbitrary permutation matrix $\mathbf{P}$.

$$\mathbf{P}^{-1} \cdot \mathbf{A} \cdot \mathbf{P} = \mathbf{A}^{\mathbf{P}} \tag{4.23}$$

4.2.3 Vectorizable formulas

Developing an FFT algorithm for SIMD processors requires replacing non-vectorizable operations with a series of vectorizable operations. This can be done by identifying formulas that can be translated into vector operations and transforming the DFT equation until it only contains vectorizable formulas.

Matrix operations on full vectors can be represented by a Kronecker product:

$$\mathbf{A} \otimes \mathbf{I}_V \tag{4.24}$$

Here, the matrix operation defined by matrix $\mathbf{A}$ is performed on vectors with V elements. In the following, the SIMD vector length is always denoted as V. For example, the matrix product defined by equation (4.25) describes a SIMD vector operation on vectors with four complex-valued elements (see figure 4.2a). On the other hand, equation (4.26) (figure 4.2b) cannot be directly vectorized. Hence, the matrix has to be rewritten using the formulas from section 4.2.2 for mapping the operation efficiently on a SIMD architecture; the rewriting introduces permutation operations defined by permutation matrices.

$$\mathbf{y} = (\mathbf{W}_2 \otimes \mathbf{I}_4) \mathbf{x} \tag{4.25}$$

$$\mathbf{y} = (\mathbf{I}_4 \otimes \mathbf{W}_2) \mathbf{x} \tag{4.26}$$

Chapter 4 Radix-2 and mixed-radix FFTs for OFDM-A and SC-FDMA

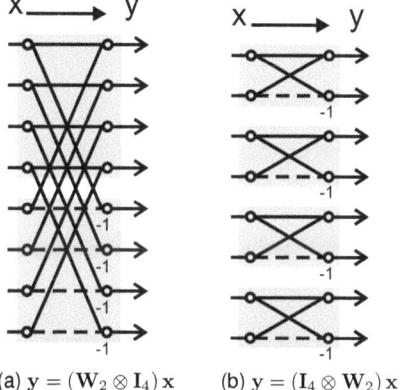

(a) $\mathbf{y} = (\mathbf{W}_2 \otimes \mathbf{I}_4)\mathbf{x}$ (b) $\mathbf{y} = (\mathbf{I}_4 \otimes \mathbf{W}_2)\mathbf{x}$

Figure 4.2: Signal flow graphs corresponding to Kronecker products with DFT matrices

Any permutation can be mapped on a series of vector permutation operations. Permutations on single vectors can be realized by a single permutation operation. The FFT algorithms mostly require permutations on pairs of vectors; examples are listed in (4.27).

$$\mathbf{P}_2^2 \otimes \mathbf{I}_{V/2}, \quad \mathbf{P}_4^2 \otimes \mathbf{I}_{V/4}, \quad \mathbf{P}_b^a \otimes \mathbf{I}_{V/ab}, \quad \mathbf{I}_{V/ab} \otimes \mathbf{P}_b^a \qquad (4.27)$$

The permutation defined by $\mathbf{P}_V^V$ represents a permutation on V vectors. This permutation can be carried out by $\log_2(V)$ permutation stages on pairs of vectors (see section 4.4.3). These permutation stages can be efficiently vectorized on any of the implemented permutation networks.
Furthermore, operations of the form $\mathbf{I}_N \otimes \mathbf{A}$ and $\mathbf{A} \cdot \mathbf{B}$ can be directly vectorized if $\mathbf{A}$ and $\mathbf{B}$ can be vectorized. For example, $\mathbf{P}_q^p \otimes \mathbf{I}_V$ performs a permutation of complete data vectors, which can be realized by adjusting the addressing of data vectors — without vector element permutations. Multiplications with complex-valued diagonal matrices, e.g. multiplications with twiddle factor matrices, can also be vectorized.

4.3 Related work on SIMD FFT algorithms

Franchetti and Püschel [FP02, FP03, Fra03, FP07, FVP07] developed FFT algorithms for general-purpose processors (GPPs) with short vector SIMD extensions, such as the MMX and SSE extensions for Intel processors, 3DNow! for AMD processors, and IBM's VMX for the Cell BE processor. The FFT algorithms are integrated into the SPIRAL code generation framework for DSP transforms [PMJ+05]. The proposed algorithms operate on real-valued data; complex-valued data is stored in an interleaved format in consecutive vector elements. Hence, every complex-valued matrix element $a = \text{Re}\{a\} + i \cdot \text{Im}\{a\}$ is replaced by a 2×2 matrix:

$$a \to \begin{bmatrix} \text{Re}\{a\} & -\text{Im}\{a\} \\ \text{Im}\{a\} & \text{Re}\{a\} \end{bmatrix} \tag{4.28}$$

In [FP03, FVP07], an FFT algorithm, based on the Cooley-Tukey FFT, for an $m \cdot n$-point DFT ($N_{\text{DFT}} = mn$) with m and n both divisible by the vector length V is proposed. The algorithm performs all DFT stages on complete vectors and requires only one vector permutation, defined by $\mathbf{P}_V^V$, for reordering elements — and two permutations for accessing real and imaginary parts of complex values (defined by $\mathbf{I}_{nm/V} \otimes \mathbf{P}_2^V$): Imaginary and real parts are stored in distinct vectors during the FFT computation and merged again after the FFT processing. Aside from the additional permutations for complex values, the algorithm has the same complexity as the mixed-radix FFT algorithm presented in section 4.4.2. Yet, the processing of real-valued data effectively doubles the vector length. Furthermore, the proposed FFT decomposition for $N_{\text{DFT}} = V \cdot M \cdot V$ is advantageous for the manual implementation of multiple FFTs, as the parts of the algorithm that are responsible for the vectorization can be reused for all DFT sizes. Both algorithms allow to interleave the $\log_2(V)$ permutation stages required for $\mathbf{P}_V^V$ with the processing stages for smaller-sized DFTs on complete vectors.

In [FP07], Franchetti and Püschel propose an FFT algorithm that vectorizes arbitrary non-power of two FFT sizes $N_{\text{DFT}} = mn$. The algorithm applies zero padding to extend both m and n to be divisible by the vector length V. The required vector element permutation operations are the same as in the algorithm for SIMD widths that are multiples of the squared vector length [FP03, FVP07]. Due to the zero padding, the algorithm does not utilize the full SIMD width, a scaling of the SIMD width does not necessarily lead to a performance gain. On a 3.6 GHz Pentium 4 processor, the speedup of an 8-way SIMD implementation compared to a scalar implementation has been measured for FFT sizes up to 100 [FP07]. For most FFT sizes, the speedup is below a factor of four.

4.4 Derivation of SIMD radix-2 and mixed-radix FFT algorithms

Below, a short radix-2 FFT algorithm is developed, which shows that radix-2 FFTs can be efficiently vectorized if the FFT size is at least twice the vector length. For mixed-radix FFTs, stricter constraints are necessary: An algorithm is proposed that enables an efficient vectorization of FFTs under the constraint that the FFT size is a multiple of the squared vector length. In this context, efficient vectorization means that all FFT stages operate on complete vectors and the number and complexity of permutation stages on vector elements is minimized.

4.4.1 Short radix-2 FFT algorithm

Vectorizing an FFT requires that all DFT stages are written as Kronecker products of the DFT matrix with $V \times V$ identity matrices (e.g. $\mathbf{W}_x \otimes \mathbf{I}_V$). Hence, the smallest radix-2 FFT size that potentially

Chapter 4 Radix-2 and mixed-radix FFTs for OFDM-A and SC-FDMA

can be vectorized in this manner is $N_{\text{DFT}} = 2 \cdot V$. A vectorized algorithm for an FFT size that is twice the vector length can be developed by repeatedly applying equation (4.5):

$$\begin{aligned}\mathbf{W}_{2 \cdot V} &= (\mathbf{W}_V \otimes \mathbf{I}_2) \cdot \mathbf{P}_V^2 \cdot \mathbf{D}_V^2 \cdot (\mathbf{W}_2 \otimes \mathbf{I}_V) \\ &= \Big(\big((\mathbf{W}_{V/2} \otimes \mathbf{I}_2) \cdot \mathbf{P}_{V/2}^2 \cdot \mathbf{D}_{V/2}^2 \cdot (\mathbf{W}_2 \otimes \mathbf{I}_{V/2})\big) \otimes \mathbf{I}_2\Big) \cdot \mathbf{P}_V^2 \cdot \mathbf{D}_V^2 \cdot (\mathbf{W}_2 \otimes \mathbf{I}_V) \\ &= (\mathbf{W}_{V/2} \otimes \mathbf{I}_4) \cdot \big(\mathbf{P}_{V/2}^2 \otimes \mathbf{I}_2\big) \cdot \big(\mathbf{D}_{V/2}^2 \otimes \mathbf{I}_2\big) \cdot (\mathbf{W}_2 \otimes \mathbf{I}_V) \cdot \mathbf{P}_V^2 \cdot \mathbf{D}_V^2 \cdot (\mathbf{W}_2 \otimes \mathbf{I}_V) \\ &= \Big(\big((\mathbf{W}_{V/4} \otimes \mathbf{I}_2) \cdot \mathbf{P}_{V/4}^2 \cdot \mathbf{D}_{V/4}^2 \cdot (\mathbf{W}_2 \otimes \mathbf{I}_{V/4})\big) \otimes \mathbf{I}_4\Big) \\ &\quad \cdot \big(\mathbf{P}_{V/2}^2 \otimes \mathbf{I}_2\big) \cdot \big(\mathbf{D}_{V/2}^2 \otimes \mathbf{I}_2\big) \cdot (\mathbf{W}_2 \otimes \mathbf{I}_V) \cdot \mathbf{P}_V^2 \cdot \mathbf{D}_V^2 \cdot (\mathbf{W}_2 \otimes \mathbf{I}_V) \\ &= (\mathbf{W}_{V/4} \otimes \mathbf{I}_8) \cdot \big(\mathbf{P}_{V/4}^2 \otimes \mathbf{I}_4\big) \cdot \big(\mathbf{D}_{V/4}^2 \otimes \mathbf{I}_4\big) \cdot (\mathbf{W}_2 \otimes \mathbf{I}_V) \\ &\quad \cdot \big(\mathbf{P}_{V/2}^2 \otimes \mathbf{I}_2\big) \cdot \big(\mathbf{D}_{V/2}^2 \otimes \mathbf{I}_2\big) \cdot (\mathbf{W}_2 \otimes \mathbf{I}_V) \\ &\quad \cdot \mathbf{P}_V^2 \cdot \mathbf{D}_V^2 \cdot (\mathbf{W}_2 \otimes \mathbf{I}_V) \\ &= \ldots \end{aligned} \qquad (4.29)$$

The decomposition in (4.29) leads to the self-sorting decimation in frequency (DIF) FFT algorithm [Tem83]. The complete algorithm may be written as:

$$\mathbf{W}_{2 \cdot V} = \mathbf{T}_0 \cdot \mathbf{T}_1 \cdots \mathbf{T}_{\log_2(V)-1} \cdot \mathbf{T}_{\log_2(V)} = \prod_{i=0}^{\log_2(V)} \mathbf{T}_i \qquad (4.30)$$

with $\quad \mathbf{T}_i = \big(\mathbf{P}_{2^i}^2 \otimes \mathbf{I}_{V/2^i}\big) \big(\mathbf{D}_{2^i}^2 \otimes \mathbf{I}_{V/2^i}\big) (\mathbf{W}_2 \otimes \mathbf{I}_V)$

In equation (4.30), all radix-2 FFT stages operate on complete vectors ($\mathbf{W}_2 \otimes \mathbf{I}_V$). The algorithm requires $\log_2(V)$ permutation stages on pairs of vectors.[1] This is the theoretical minimum number of permutation stages, as every element of each output vector of the DFT depends on every element of each input vector; the necessary reordering requires at least $\log_2(V)$ permutation operations.

Figure 4.3 displays the signal-flow graph of an 8-point vectorized self-sorting DIF FFT for a vector length of four. The matrix representation is shown in equation (4.31). The different stages of the algorithm are marked by Roman numbers in the equation and the figure; the signals flow from left to right. Data vectors have also been highlighted by light gray boxes in the radix-2 FFT stages.

$$\mathbf{W}_8 = \underbrace{(\mathbf{W}_2 \otimes \mathbf{I}_4)}_{\text{VII}} \cdot \underbrace{(\mathbf{P}_2^2 \otimes \mathbf{I}_2)}_{\text{VI}} \cdot \underbrace{(\mathbf{D}_2^2 \otimes \mathbf{I}_2)}_{\text{V}} \cdot \underbrace{(\mathbf{W}_2 \otimes \mathbf{I}_4)}_{\text{IV}} \cdot \underbrace{\mathbf{P}_4^2}_{\text{III}} \cdot \underbrace{\mathbf{D}_4^2}_{\text{II}} \cdot \underbrace{(\mathbf{W}_2 \otimes \mathbf{I}_4)}_{\text{I}} \qquad (4.31)$$

Extension to longer radix-2 FFTs

An extension of the FFT algorithm for $N_{\text{DFT}} = 2 \cdot V$ to longer radix-2 FFTs can be done by adding further factors to the FFT and applying equation (4.5) to decompose the FFT. Equation (4.32) shows the algorithm for an N-point radix-2 FFT for $N \geq 2 \cdot V$. The principle structure of the algorithm stays the same. The number of permutation stages increases, yet the number and complexity of

---
[1] The permutation stage for $i = 0$ degenerates to a product with an identity matrix: $\mathbf{P}_1^2 \otimes \mathbf{I}_V = \mathbf{I}_{2 \cdot V}$.

Chapter 4 Radix-2 and mixed-radix FFTs for OFDM-A and SC-FDMA

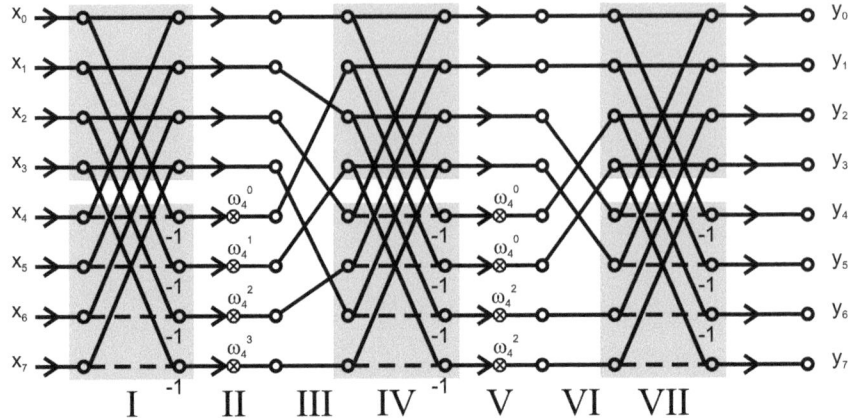

Figure 4.3: Signal-flow graph for a vectorized 8-point FFT for a SIMD width of four complex-valued elements.

permutation operations that require vector element permutations is the same as before; all additional permutation stages perform permutations on complete vectors.

$$\mathbf{W}_N = \prod_{i=0}^{\log_2(N)-1} \left(\mathbf{P}_{2^i}^2 \otimes \mathbf{I}_{N/2^{i+1}} \right) \left(\mathbf{D}_{2^i}^2 \otimes \mathbf{I}_{N/2^{i+1}} \right) \left(\mathbf{W}_2 \otimes \mathbf{I}_{N/2} \right) \tag{4.32}$$

Using equation (4.32), any power of two DFT, whose size is at least twice the vector length, can be vectorized with all radix-2 FFT stages operating on complete vectors and exactly $\log_2(V)$ permutation stages on pairs of vectors for the reordering of elements during the FFT.

Vectorization of short mixed-radix FFTs

An extension of the FFT algorithm in equation (4.30) for $N_{\text{DFT}} = 2 \cdot V$ to the mixed-radix case can be done by adding a factor m that contains all non-power of two factors (e.g. $m = 3^a \cdot 5^b$). Using the basic FFT decomposition formula (4.5), the following FFT algorithm can be derived:

$$\begin{aligned}
\mathbf{W}_{2 \cdot V \cdot m} &= (\mathbf{W}_{2 \cdot V} \otimes \mathbf{I}_m) \cdot \mathbf{P}_{2 \cdot V}^m \cdot \mathbf{D}_{2 \cdot V}^m \cdot (\mathbf{W}_m \otimes \mathbf{I}_{2 \cdot V}) \\
&= \left(\left(\prod_{i=0}^{\log_2(V)} \mathbf{T}_i \right) \otimes \mathbf{I}_m \right) \cdot \mathbf{P}_{2 \cdot V}^m \cdot \mathbf{D}_{2 \cdot V}^m \cdot (\mathbf{W}_m \otimes \mathbf{I}_{2 \cdot V}) \\
&= \mathbf{P}_{2 \cdot V}^m \cdot \left(\left(\mathbf{I}_m \otimes \prod_{i=0}^{\log_2(V)} \mathbf{T}_i \right) \right) \cdot \mathbf{D}_{2 \cdot V}^m \cdot (\mathbf{W}_m \otimes \mathbf{I}_{2 \cdot V})
\end{aligned} \tag{4.33}$$

$$\mathbf{W}_{2\cdot V\cdot m} = \mathbf{P}_{2\cdot V}^m \cdot \left(\prod_{i=0}^{\log_2(V)} \left(\mathbf{I}_m \otimes \mathbf{P}_{2^i}^2 \otimes \mathbf{I}_{V/2^i} \right) \left(\mathbf{I}_m \otimes \mathbf{D}_{2^i}^2 \otimes \mathbf{I}_{V/2^i} \right) \left(\mathbf{I}_m \otimes \mathbf{W}_2 \otimes \mathbf{I}_V \right) \right)$$
$$\cdot \mathbf{D}_{2\cdot V}^m \cdot \left(\mathbf{W}_m \otimes \mathbf{I}_{2\cdot V} \right) \tag{4.34}$$

A closer look at equation (4.34) shows that all radix-2 as well as the m-point FFT stage operate on complete vectors (vector elements at a distance of m vectors for the radix-2 stages and vector elements at a distance of two vectors for the m-point FFT stage), independent of the further decomposition of the m-point FFT into smaller DFTs. Yet, the additional factor m introduces a vector element permutation defined by $\mathbf{P}_{2\cdot V}^m$.

The permutation matrix $\mathbf{P}_{2\cdot V}^m$ defines a stride by $2 \cdot V$ permutation on $2 \cdot V \cdot m$ vectors. Figure 4.4 depicts an example for $m = 3$ and $V = 4$; the permutation input has a length of six vectors. In the example, each output vector contains values from three input vectors; hence, $\mathbf{P}_8^3$ can be realized by two separate permutation stages with three input and output vectors each. In general, the complexity of $\mathbf{P}_{2\cdot V}^m$ depends on m and the ratio of m to the vector length V.

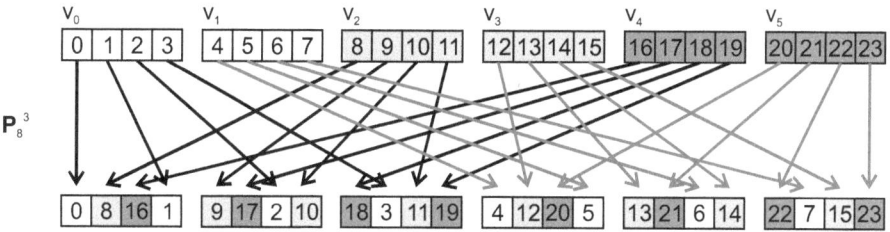

Figure 4.4: Stride permutation defined by $\mathbf{P}_8^3$ for a vector length of four elements

The permutation defined by $\mathbf{P}_{2\cdot V}^m$ is the last processing stage of the algorithm and cannot be interleaved with any of the previous processing stages to hide the permutation overhead. Moving the permutation to the right of the processing step for $\mathbf{W}_{2\cdot V}$ would enable an interleaving of DFT stages and permutation stages, yet moving $\mathbf{P}_{2\cdot V}^m$ also increases the complexity of permutation stages during the $2 \cdot V$-point FFT, as can be seen below:

$$\mathbf{P}_{2\cdot V}^m \cdot \left(\prod_{i=0}^{\log_2(V)} \left(\mathbf{I}_m \otimes \mathbf{P}_{2^i}^2 \otimes \mathbf{I}_{V/2^i} \right) \left(\mathbf{I}_m \otimes \mathbf{D}_{2^i}^2 \otimes \mathbf{I}_{V/2^i} \right) \left(\mathbf{I}_m \otimes \mathbf{W}_2 \otimes \mathbf{I}_V \right) \right)$$
$$= \left(\prod_{i=0}^{\log_2(V)} \left(\mathbf{P}_{2^i}^2 \otimes \mathbf{I}_{m\cdot V/2^i} \right) \left(\mathbf{D}_{2^i}^2 \otimes \mathbf{I}_{m\cdot V/2^i} \right) \left(\mathbf{W}_2 \otimes \mathbf{I}_{m\cdot V} \right) \right) \cdot \mathbf{P}_{2\cdot V}^m \tag{4.35}$$

The permutations defined by $\mathbf{P}_{2^i}^2 \otimes \mathbf{I}_{m\cdot V/2^i}$ during the $2 \cdot V$-point FFT in equation (4.35) are more complex than permutations in the radix-2 case. In the radix-2 case ($\mathbf{P}_{2^i}^2 \otimes \mathbf{I}_{V/2^i}$), permutations are performed on blocks of 1, 2, 4, ..., $V/2$ vector elements ($\mathbf{P}_V^2, \mathbf{P}_{V/2}^2 \otimes \mathbf{I}_2, \mathbf{P}_{V/4}^2 \otimes \mathbf{I}_4, \ldots, \mathbf{P}_2^2 \otimes \mathbf{I}_{V/2}$). The vector length is always a multiple of the block size, thus the alignment on data vectors is preserved.

In the mixed-radix case ($\mathbf{P}_{2^i}^2 \otimes \mathbf{I}_{m \cdot V/2^i}$), permutations are performed on blocks of $m, 2 \cdot m, 4 \cdot m, \ldots,$ $m \cdot V/2$ vector elements ($\mathbf{P}_V^2 \otimes \mathbf{I}_m, \mathbf{P}_{V/2}^2 \otimes \mathbf{I}_{2 \cdot m}, \mathbf{P}_{V/4}^2 \otimes \mathbf{I}_{4 \cdot m}, \ldots, \mathbf{P}_2^2 \otimes \mathbf{I}_{m \cdot V/2}$). These permutations do not preserve the alignment on vectors, as can be seen by the examples in figure 4.5 for $m = 3$ and $V = 4$. The first permutation stage (figure 4.5a), defined by $\mathbf{P}_4^2 \otimes \mathbf{I}_3$, operates on blocks of three vector elements. The permuted blocks have to be stored across data vectors. The same effect occurs for the second permutation stage (figure 4.5b), defined by $\mathbf{P}_2^2 \otimes \mathbf{I}_6$.

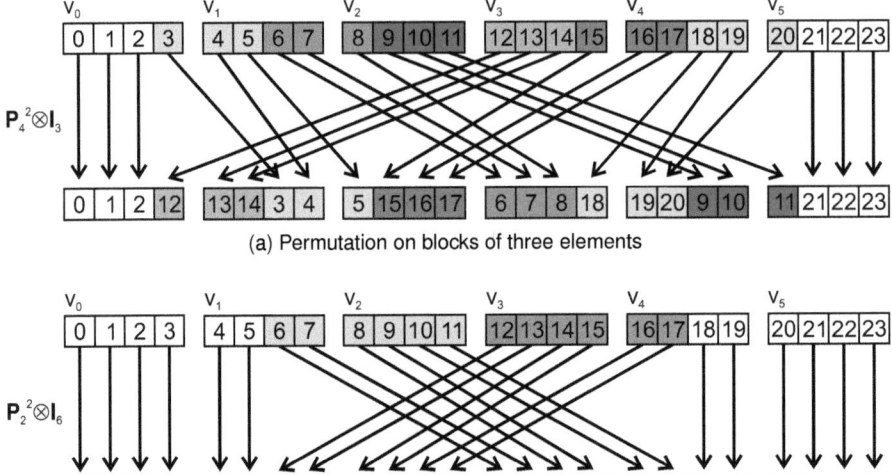

Figure 4.5: Permutations in between radix-2 FFT stages for the mixed-radix FFT with $m = 3$ and $V = 4$: (a) $\mathbf{P}_4^2 \otimes \mathbf{I}_3$ and (b) $\mathbf{P}_2^2 \otimes \mathbf{I}_6$

Another reason why the short mixed-radix FFT algorithm has a higher complexity than the short radix-2 FFT algorithm is the implementation of permutations on the different permutation networks of the proposed scalable SIMD architecture. Permutations for the radix-2 FFT can mostly be mapped on simple masked butterfly permutations on pairs of vectors (see section 4.4.3), yet this is not possible for $\mathbf{P}_{2 \cdot V}^m$. Hence, the mixed-radix FFT requires a more restrictive constraint on the ratio between SIMD vector length V and FFT length than the radix-2 FFT. In the radix-2 case, the constraint that the FFT size should be at least *twice the vector length* is sufficient.

4.4.2 Mixed-radix FFT algorithm

The mixed-radix algorithm in equation (4.34) for $N_{\text{DFT}} = 2 \cdot V \cdot m$ does not enable efficient vector implementations for arbitrary m, due to the complexity of the permutation stage defined by $\mathbf{P}_{2 \cdot V}^m$. If m is a multiple of the vector length V (i.e. $m = l \cdot V$), the complexity of the vector element permutation is

Chapter 4 Radix-2 and mixed-radix FFTs for OFDM-A and SC-FDMA

no longer dependent on the non-power of two factors in m. This can be shown by applying equation (4.17):

$$\mathbf{P}_{2 \cdot V}^{l \cdot V} = \left(\mathbf{P}_{2 \cdot V}^{l} \otimes \mathbf{I}_V\right) \cdot \left(\mathbf{I}_l \otimes \mathbf{P}_{2 \cdot V}^{V}\right) \tag{4.36}$$

Therefore, the constraint that the FFT size should be a *multiple of the squared vector length* could be a sufficient constraint for the vectorization of mixed-radix FFTs. In the following, an FFT algorithm is developed, which evidences that an efficient vectorization of mixed-radix FFTs is indeed possible if the FFT size fulfills this constraint.

Mixed-radix FFT algorithm derivation

The proposed mixed-radix FFT algorithm is designed for an FFT size $N_{\text{DFT}} = V \cdot M \cdot V$. Here, the factor M contains all non-power of two factors. The DFT matrix $\mathbf{W}_{V \cdot M \cdot V}$ is first decomposed into three factors by applying the basic FFT decomposition in equation (4.5) twice:

$$\begin{aligned}
\mathbf{W}_{VMV} &= (\mathbf{W}_{VM} \otimes \mathbf{I}_V) \cdot \mathbf{P}_{MV}^{V} \cdot \mathbf{D}_{MV}^{V} \cdot (\mathbf{W}_V \otimes \mathbf{I}_{VM}) \\
&= \left(((\mathbf{W}_V \otimes \mathbf{I}_M) \cdot \mathbf{P}_V^M \cdot \mathbf{D}_V^M \cdot (\mathbf{W}_M \otimes \mathbf{I}_V)) \otimes \mathbf{I}_V\right) \\
&\quad \cdot \mathbf{P}_{MV}^{V} \cdot \mathbf{D}_{MV}^{V} \cdot (\mathbf{W}_V \otimes \mathbf{I}_{VM}) \\
&= (\mathbf{W}_V \otimes \mathbf{I}_M \otimes \mathbf{I}_V) \cdot (\mathbf{P}_V^M \otimes \mathbf{I}_V) \cdot (\mathbf{D}_V^M \otimes \mathbf{I}_V) \cdot (\mathbf{W}_M \otimes \mathbf{I}_V \otimes \mathbf{I}_V) \\
&\quad \cdot \mathbf{P}_{MV}^{V} \cdot \mathbf{D}_{MV}^{V} \cdot (\mathbf{W}_V \otimes \mathbf{I}_{VM})
\end{aligned} \tag{4.37}$$

Next, the permutation stages can be reordered by applying formula manipulations as defined in section 4.2.2. The permutation defined by $\mathbf{P}_V^M \otimes \mathbf{I}_V$ can be moved to the left by first applying formula (4.14) and then formula (4.16).

$$\begin{aligned}
\mathbf{W}_{VMV} &= \left(((\mathbf{W}_V \otimes \mathbf{I}_M) \cdot \mathbf{P}_V^M) \otimes \mathbf{I}_V\right) \cdot (\mathbf{D}_V^M \otimes \mathbf{I}_V) \cdot (\mathbf{W}_M \otimes \mathbf{I}_V \otimes \mathbf{I}_V) \\
&\quad \cdot \mathbf{P}_{MV}^{V} \cdot \mathbf{D}_{MV}^{V} \cdot (\mathbf{W}_V \otimes \mathbf{I}_{VM}) \\
&= (\mathbf{P}_V^M \otimes \mathbf{I}_V) \cdot (\mathbf{I}_M \otimes \mathbf{W}_V \otimes \mathbf{I}_V) \cdot (\mathbf{D}_V^M \otimes \mathbf{I}_V) \cdot (\mathbf{W}_M \otimes \mathbf{I}_V \otimes \mathbf{I}_V) \\
&\quad \cdot \mathbf{P}_{MV}^{V} \cdot \mathbf{D}_{MV}^{V} \cdot (\mathbf{W}_V \otimes \mathbf{I}_{VM})
\end{aligned} \tag{4.38}$$

Permutation $\mathbf{P}_{MV}^{V}$ can also be moved to the left using formula (4.16). The result can be represented by three consecutive matrix operations $\mathbf{T}_{V_1}$, $\mathbf{T}_M$, and $\mathbf{T}_{V_2}$:

$$\begin{aligned}
\mathbf{W}_{VMV} &= (\mathbf{P}_V^M \otimes \mathbf{I}_V) \cdot (\mathbf{I}_M \otimes \mathbf{W}_V \otimes \mathbf{I}_V) \cdot (\mathbf{D}_V^M \otimes \mathbf{I}_V) \cdot \mathbf{P}_{MV}^{V} \cdot (\mathbf{I}_V \otimes \mathbf{W}_M \otimes \mathbf{I}_V) \\
&\quad \cdot \mathbf{D}_{MV}^{V} \cdot (\mathbf{W}_V \otimes \mathbf{I}_{VM})
\end{aligned} \tag{4.39}$$

$$\mathbf{W}_{VMV} = (\mathbf{P}_V^M \otimes \mathbf{I}_V) \cdot (\mathbf{I}_M \otimes \mathbf{W}_V \otimes \mathbf{I}_V) \cdot \mathbf{P}_{MV}^{V} \cdot (\mathbf{I}_V \otimes \mathbf{D}_V^M) \cdot (\mathbf{I}_V \otimes \mathbf{W}_M \otimes \mathbf{I}_V) \tag{4.40}$$
$$\quad \cdot \mathbf{D}_{MV}^{V} \cdot (\mathbf{W}_V \otimes \mathbf{I}_{VM})$$

$$\mathbf{W}_{VMV} = \mathbf{T}_{V_1} \cdot \mathbf{T}_M \cdot \mathbf{T}_{V_2} \tag{4.41}$$

$$\mathbf{T}_{V_1} = (\mathbf{P}_V^M \otimes \mathbf{I}_V) \cdot (\mathbf{I}_M \otimes \mathbf{W}_V \otimes \mathbf{I}_V) \cdot \mathbf{P}_{MV}^{V} \tag{4.42}$$

$$\mathbf{T}_M = (\mathbf{I}_V \otimes \mathbf{D}_V^M) \cdot (\mathbf{I}_V \otimes \mathbf{W}_M \otimes \mathbf{I}_V) \tag{4.43}$$

Chapter 4 Radix-2 and mixed-radix FFTs for OFDM-A and SC-FDMA

$$\mathbf{T}_{V_2} = \mathbf{D}_{MV}^V \cdot (\mathbf{W}_V \otimes \mathbf{I}_{VM}) \tag{4.44}$$

The matrix operations defined by $\mathbf{T}_{V_2}$ and $\mathbf{T}_M$ (equations (4.43) and (4.44)) can be directly mapped on vector operations: All DFT stages operate on complete vectors, as evidenced by the Kronecker products with unity matrices of length V. The multiplications by twiddle factor matrices are also vectorizable, as the twiddle factor matrices are diagonal matrices.

Formula $\mathbf{T}_{V_1}$ in equation (4.42) contains the permutation matrix $\mathbf{P}_{MV}^V$, which cannot be directly vectorized. Hence, further formula manipulations are necessary. Using formula (4.18), permutation matrix $\mathbf{P}_{MV}^V$ can be split into a pair of permutations. Next, the matrices are reordered using formula (4.15):

$$\begin{aligned}
\mathbf{T}_{V_1} &= \left(\mathbf{P}_V^M \otimes \mathbf{I}_V\right) \cdot \left(\mathbf{I}_M \otimes \mathbf{W}_V \otimes \mathbf{I}_V\right) \cdot \mathbf{P}_{MV}^V \\
&= \left(\mathbf{P}_V^M \otimes \mathbf{I}_V\right) \cdot \left(\mathbf{I}_M \otimes \mathbf{W}_V \otimes \mathbf{I}_V\right) \cdot \underbrace{\left(\mathbf{I}_M \otimes \mathbf{P}_V^V\right) \cdot \left(\mathbf{P}_M^V \otimes \mathbf{I}_V\right)}_{\mathbf{P}_{MV}^V} \\
&= \left(\mathbf{P}_V^M \otimes \mathbf{I}_V\right) \cdot \left(\mathbf{I}_M \otimes \left((\mathbf{W}_V \otimes \mathbf{I}_V) \cdot \mathbf{P}_V^V\right)\right) \cdot \left(\mathbf{P}_M^V \otimes \mathbf{I}_V\right)
\end{aligned} \tag{4.45}$$

Permutation $\mathbf{P}_M^V \otimes \mathbf{I}_V$ performs a reordering of complete data vectors, which is later reversed by $\mathbf{P}_V^M \otimes \mathbf{I}_V$. This can be shown by applying formulas (4.19) and (4.20); the formula can be simplified using (4.23).

$$\begin{aligned}
\mathbf{T}_{V_1} &= \left(\left(\mathbf{P}_M^V \otimes \mathbf{I}_V\right)^{-1}\right)^{-1} \cdot \left(\mathbf{I}_M \otimes \left((\mathbf{W}_V \otimes \mathbf{I}_V) \cdot \mathbf{P}_V^V\right)\right) \cdot \left(\mathbf{P}_M^V \otimes \mathbf{I}_V\right) \\
&= \left(\mathbf{P}_M^V \otimes \mathbf{I}_V\right)^{-1} \cdot \left(\mathbf{I}_M \otimes \left((\mathbf{W}_V \otimes \mathbf{I}_V) \cdot \mathbf{P}_V^V\right)\right) \cdot \left(\mathbf{P}_M^V \otimes \mathbf{I}_V\right) \\
&= \left(\mathbf{I}_M \otimes \left((\mathbf{W}_V \otimes \mathbf{I}_V) \cdot \mathbf{P}_V^V\right)\right)^{\mathbf{P}_M^V \otimes \mathbf{I}_V}
\end{aligned} \tag{4.46}$$

Matrix $\mathbf{T}_{V_2}$ can be rewritten in a similar manner by expanding with $\left(\mathbf{P}_V^M \otimes \mathbf{I}_V\right) \cdot \left(\mathbf{P}_M^V \otimes \mathbf{I}_V\right)$ and applying formulas (4.16), (4.20) and (4.23).

$$\begin{aligned}
\mathbf{T}_{V_2} &= \mathbf{D}_{MV}^V \cdot (\mathbf{W}_V \otimes \mathbf{I}_{VM}) \\
&= \mathbf{D}_{MV}^V \cdot (\mathbf{W}_V \otimes \mathbf{I}_{VM}) \underbrace{\left(\mathbf{P}_V^M \otimes \mathbf{I}_V\right) \cdot \left(\mathbf{P}_M^V \otimes \mathbf{I}_V\right)}_{\mathbf{I}_{VMV}} \\
&= \mathbf{D}_{MV}^V \cdot \left(\mathbf{P}_V^M \otimes \mathbf{I}_V\right) \cdot \left(\mathbf{I}_M \otimes \mathbf{W}_V \otimes \mathbf{I}_V\right) \cdot \left(\mathbf{P}_M^V \otimes \mathbf{I}_V\right) \\
&= \mathbf{D}_{MV}^V \cdot \left(\mathbf{I}_M \otimes \mathbf{W}_V \otimes \mathbf{I}_V\right)^{\mathbf{P}_M^V \otimes \mathbf{I}_V}
\end{aligned} \tag{4.47}$$

Using formulas (4.47), (4.43), and (4.46), the complete FFT algorithm may be written as follows:

$$\begin{aligned}
\mathbf{W}_{VMV} = &\left(\mathbf{I}_M \otimes \left((\mathbf{W}_V \otimes \mathbf{I}_V) \cdot \mathbf{P}_V^V\right)\right)^{\mathbf{P}_M^V \otimes \mathbf{I}_V} \\
&\cdot \left(\mathbf{I}_V \otimes \mathbf{D}_V^M\right) \cdot \left(\mathbf{I}_V \otimes \mathbf{W}_M \otimes \mathbf{I}_V\right) \\
&\cdot \mathbf{D}_{MV}^V \cdot \left(\mathbf{I}_M \otimes \mathbf{W}_V \otimes \mathbf{I}_V\right)^{\mathbf{P}_M^V \otimes \mathbf{I}_V}
\end{aligned} \tag{4.48}$$

Mixed-radix FFT algorithm properties

The FFT algorithm defined by equation (4.48) can be efficiently vectorized, independent of the further decomposition of V-point and M-point DFTs. All FFT stages ($\mathbf{W}_V \otimes \mathbf{I}_V$ and $\mathbf{W}_M \otimes \mathbf{I}_V$) operate on complete data vectors. Furthermore, the algorithm contains only one permutation on vector elements, defined by $\mathbf{P}_V^V$. The permutation $\mathbf{P}_V^V$ does not depend on M, which contains the non-power of two factors, and can be carried out in $\log_2(V)$ permutation stages on pairs of vectors (see section 4.4.3). The number of permutation stages is the same as for the radix-2 FFT algorithm. If the V-point DFT in $\mathbf{T}_{V_1}$ (equation (4.46)) is further decomposed into a series of $\log_2(V)$ radix-2 butterfly stages, the permutation stages can be interleaved with the radix-2 butterfly stages. The decomposition of permutations and the interleaving with radix-2 butterfly stages are explained in detail in section 4.4.3. The FFT algorithm consists of two parts, a pair of V-point DFTs, defined by $\mathbf{T}_{V_1}$ and $\mathbf{T}_{V_2}$, and an M-point DFT, defined by $\mathbf{T}_M$. Both $\mathbf{T}_{V_1}$ and $\mathbf{T}_{V_2}$ share a similar structure. The only differences are that $\mathbf{T}_{V_1}$ contains a permutation by matrix $\mathbf{P}_V^V$ and $\mathbf{T}_{V_2}$ comprises a twiddle factor multiplication (with $\mathbf{D}_{MV}^V$). The actual processing of the V-point FFTs is enclosed by a similarity transformation by $\mathbf{P}_M^V \otimes \mathbf{I}_V$ (see formula (4.23)). This transformation describes the addressing of data vectors: Every M-th data vector is used as input for one V-point FFT on data vectors, e.g. vectors $0, M, 2 \cdot M, \ldots, (V-1) \cdot M$ are the input vectors for the first V-point FFT. The output of the V-point FFT is stored at the same positions (every M-th vector). The Kronecker product $\mathbf{I}_M \otimes \ldots$ means that M V-point FFTs have to be carried out in this manner on different input vectors. Figure 4.6 illustrates the data flow between V-point and M-point FFT stages by an example for $M = 3$ and $V = 4$.

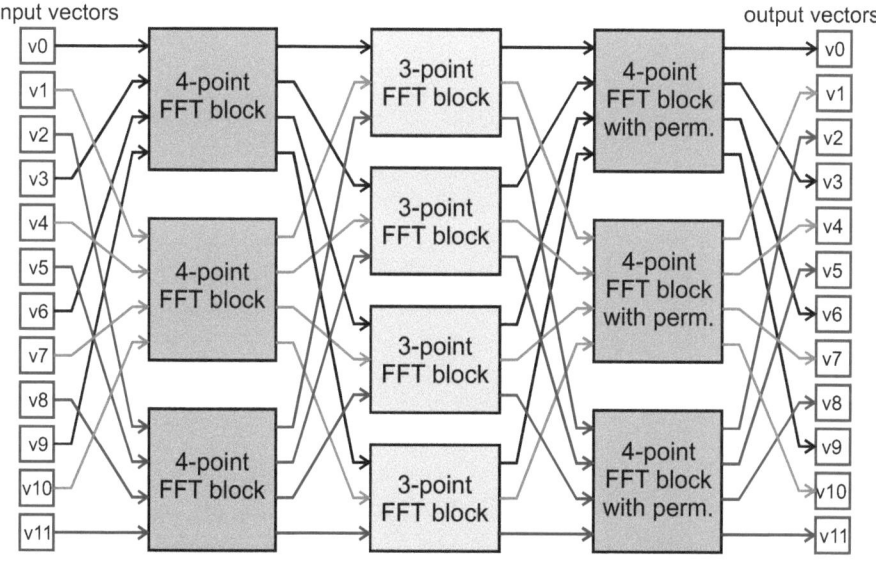

Figure 4.6: Block diagram of a 48-point mixed-radix FFT for a vector length of four

The matrix operations defined by $\mathbf{T}_{V_1}$ and $\mathbf{T}_{V_2}$ only depend on the factor M to a minor degree. M influences the values (and the length) of the twiddle factor matrix $\mathbf{D}_{MV}^{V}$, the addressing of data vectors (described by $\mathbf{P}_{M}^{V} \otimes \mathbf{I}_{V}$), and the number of V-point radix-2 FFTs. The twiddle factor matrix can be automatically generated for any value of M, the additional effort if M changes is negligible. From an implementation point of view, the addressing of data vectors and the number of FFTs can be adjusted by parametrizing pointer increments and loop iteration counts with M. Therefore, both $\mathbf{T}_{V_1}$ and $\mathbf{T}_{V_2}$ can be computed, implemented, and optimized once and later reused for arbitrary M. This is a major advantage in comparison to the FFT algorithm by Franchetti and Püschel [FP03, FVP07]. The M-point DFT, defined by $\mathbf{T}_M$, performs V FFTs, each operating on M consecutive data vectors. All operations are operations on complete data vectors. Hence, any algorithm can be applied to decompose $\mathbf{W}_M$ into smaller DFTs. Furthermore, M may take on arbitrary values; therefore, arbitrary DFTs can be vectorized as long as the constraint that the DFT size is a multiple of the squared SIMD vector length is satisfied.

4.4.3 Permutations for the vectorized FFT algorithms

This section consists of two parts. Firstly, it is proven that the permutation $\mathbf{P}_{V}^{V}$, which is required for the mixed-radix FFT (with $N_{\text{DFT}} = V \cdot M \cdot V$), can be decomposed into the same stride permutations used for the radix-2 FFT (with $N_{\text{DFT}} \geq 2 \cdot V$) and some additional permutations, which permute entire vectors. The interleaving of the required permutations with the radix-2 FFT stages is also demonstrated. Secondly, permutation operations based on butterfly permutations on pairs of vectors are explained. These permutation operations are better suited for single-vector permutation networks than stride permutations and can be realized by pre-defined butterfly permutation instructions on all implemented permutation networks.

Decomposition of permutations for the mixed-radix FFT

The proposed radix-2 FFT algorithm for $N_{\text{DFT}} = 2 \cdot V$ (equation (4.30)) requires $\log_2(V)$ permutation operations that reorder vector elements. These permutation stages operate on pairs of vectors and can be written as:

$$\mathbf{P}_{2^i}^{2} \otimes \mathbf{I}_{V/2^i} \qquad \forall i \in [1, \log_2(V)] \tag{4.49}$$

The permutation $\mathbf{P}_{V}^{V}$, which is required for the mixed-radix FFT performs exactly the same permutations on vector elements (and some permutations on complete vectors). This can be proven by decomposing $\mathbf{P}_{V}^{V}$ into smaller factors using formula (4.17) and then formula (4.18). For a vector length of four, the permutation may be decomposed as:

$$\begin{aligned}\mathbf{P}_{4}^{4} &= \left(\mathbf{P}_{4}^{2} \otimes \mathbf{I}_{2}\right) \cdot \left(\mathbf{I}_{2} \otimes \mathbf{P}_{4}^{2}\right) \\ &= \underbrace{\left(\mathbf{I}_{2} \otimes \mathbf{P}_{2}^{2} \otimes \mathbf{I}_{2}\right)}_{\mathbf{A}_1} \cdot \underbrace{\left(\mathbf{P}_{2}^{2} \otimes \mathbf{I}_{4}\right)}_{\mathbf{B}_1} \cdot \underbrace{\left(\mathbf{I}_{2} \otimes \mathbf{P}_{4}^{2}\right)}_{\mathbf{C}_1}\end{aligned} \tag{4.50}$$

The permutation defined by $\mathbf{B}_1$ performs a permutation on complete vectors; matrices $\mathbf{A}_1$ and $\mathbf{C}_1$ perform exactly the same permutations as required for the radix-2 FFT. If the vector length is doubled

($V = 8$), the number of permutation stages on vector elements increases to three, as in the radix-2 case. The decomposition is also done by first applying formula (4.17) and then formula (4.18).

$$\begin{aligned}\mathbf{P}_8^8 &= \left(\mathbf{P}_8^4 \otimes \mathbf{I}_2\right) \cdot \left(\mathbf{I}_4 \otimes \mathbf{P}_8^2\right) \\ &= \underbrace{\left(\mathbf{I}_2 \otimes \mathbf{P}_4^4 \otimes \mathbf{I}_2\right)}_{A_2} \cdot \underbrace{\left(\mathbf{P}_2^4 \otimes \mathbf{I}_8\right)}_{B_2} \cdot \underbrace{\left(\mathbf{I}_2 \otimes \mathbf{P}_8^2\right)}_{C_2}\end{aligned} \quad (4.51)$$

Here, permutation B_2 is again a permutation on complete vectors. Matrix C_2 defines the first permutation stage on vector elements and A_2 defines the remaining two permutation stages, which can be computed by extending the permutations defined in (4.50) by a Kronecker product with a 2×2 identity matrix. In the same manner, the permutation stages for an arbitrary vector length V can be computed recursively from the permutations for $\mathbf{P}_{V/2}^{V/2}$:

$$\begin{aligned}\mathbf{P}_V^V &= \left(\mathbf{P}_V^{V/2} \otimes \mathbf{I}_2\right) \cdot \left(\mathbf{I}_{V/2} \otimes \mathbf{P}_V^2\right) \\ &= \left(\mathbf{I}_2 \otimes \mathbf{P}_{V/2}^{V/2} \otimes \mathbf{I}_2\right) \cdot \left(\mathbf{P}_2^{V/2} \otimes \mathbf{I}_V\right) \cdot \left(\mathbf{I}_2 \otimes \mathbf{P}_V^2\right)\end{aligned} \quad (4.52)$$

The permutation stages required for $\mathbf{P}_V^V$ can also be efficiently interleaved with the processing of a V-point FFT, i. e. each permutation stage on pairs of vectors is followed by a radix-2 FFT stage on the same pairs of vectors. This enables a better scheduling of instructions on the processor. The principle approach is explained in the following. First, the V-point DFT matrix and the permutation matrix are decomposed into smaller factors using formula (4.5) and formula (4.17). Next, the permutation defined by $\mathbf{P}_{V/2}^2 \otimes \mathbf{I}_V$ is moved to the left using formula (4.16). The same formula is then applied for interleaving $\mathbf{P}_V^{V/2} \otimes \mathbf{I}_2$ with the two DFT operations.

$$\begin{aligned}(\mathbf{W}_V \otimes \mathbf{I}_V) \cdot \mathbf{P}_V^V &= \underbrace{\left(\mathbf{W}_{V/2} \otimes \mathbf{I}_{2 \cdot V}\right) \cdot \left(\mathbf{P}_{V/2}^2 \otimes \mathbf{I}_V\right) \cdot \left(\mathbf{D}_{V/2}^2 \otimes \mathbf{I}_V\right) \cdot \left(\mathbf{W}_2 \otimes \mathbf{I}_{V \cdot V/2}\right)}_{\mathbf{W}_V \otimes \mathbf{I}_V} \\ &\quad \cdot \underbrace{\left(\mathbf{P}_V^{V/2} \otimes \mathbf{I}_2\right) \cdot \left(\mathbf{I}_{V/2} \otimes \mathbf{P}_V^2\right)}_{\mathbf{P}_V^V} \\ &= \left(\mathbf{P}_{V/2}^2 \otimes \mathbf{I}_V\right) \cdot \left(\mathbf{I}_2 \otimes \mathbf{W}_{V/2} \otimes \mathbf{I}_V\right) \cdot \left(\mathbf{D}_{V/2}^2 \otimes \mathbf{I}_V\right) \\ &\quad \cdot \left(\mathbf{W}_2 \otimes \mathbf{I}_{V \cdot V/2}\right) \cdot \left(\mathbf{P}_V^{V/2} \otimes \mathbf{I}_2\right) \cdot \left(\mathbf{I}_{V/2} \otimes \mathbf{P}_V^2\right) \\ &= \left(\mathbf{P}_{V/2}^2 \otimes \mathbf{I}_V\right) \cdot \underbrace{\left(\mathbf{I}_2 \otimes \mathbf{W}_{V/2} \otimes \mathbf{I}_V\right) \cdot \left(\mathbf{P}_V^{V/2} \otimes \mathbf{I}_2\right)}_{\Omega} \\ &\quad \cdot \left(\mathbf{I}_{V/2} \otimes \mathbf{D}_{V/2}^2 \otimes \mathbf{I}_2\right) \cdot \left(\mathbf{I}_{V/2} \otimes \mathbf{W}_2 \otimes \mathbf{I}_V\right) \cdot \left(\mathbf{I}_{V/2} \otimes \mathbf{P}_V^2\right) \\ &= \left(\mathbf{P}_{V/2}^2 \otimes \mathbf{I}_V\right) \cdot \Omega \cdot \left(\mathbf{I}_{V/2} \otimes \mathbf{D}_{V/2}^2 \otimes \mathbf{I}_2\right) \cdot \left(\mathbf{I}_{V/2} \otimes \mathbf{W}_2 \otimes \mathbf{I}_V\right) \cdot \left(\mathbf{I}_{V/2} \otimes \mathbf{P}_V^2\right)\end{aligned} \quad (4.53)$$

The matrix Ω requires further processing to prove that the permutation stages on vector elements can always be interleaved with the processing of smaller DFTs. The permutation matrix $\mathbf{P}_V^{V/2}$ can be decomposed into two permutations using formula (4.18). Next, the matrices can be reordered by applying formula (4.14).

$$\begin{aligned}
\Omega &= \left(\mathbf{I}_2 \otimes \mathbf{W}_{V/2} \otimes \mathbf{I}_V\right) \cdot \left(\mathbf{P}_V^{V/2} \otimes \mathbf{I}_2\right) \\
&= \left(\mathbf{I}_2 \otimes \mathbf{W}_{V/2} \otimes \mathbf{I}_V\right) \cdot \underbrace{\left(\mathbf{I}_2 \otimes \mathbf{P}_{V/2}^{V/2} \otimes \mathbf{I}_2\right) \cdot \left(\mathbf{P}_2^{V/2} \otimes \mathbf{I}_V\right)}_{\mathbf{P}_V^{V/2} \otimes \mathbf{I}_2} \\
&= \left(\mathbf{I}_2 \otimes \left((\mathbf{W}_{V/2} \otimes \mathbf{I}_{V/2}) \cdot \mathbf{P}_{V/2}^{V/2}\right) \otimes \mathbf{I}_2\right) \cdot \left(\mathbf{P}_2^{V/2} \otimes \mathbf{I}_V\right)
\end{aligned} \quad (4.54)$$

Applying Ω to formula (4.53) leads to equation (4.55). The equation has been simplified by formula (4.23), as $\left(\mathbf{P}_{V/2}^2 \otimes \mathbf{I}_V\right)$ is the inverse permutation of $\left(\mathbf{P}_2^{V/2} \otimes \mathbf{I}_V\right)$.

$$\begin{aligned}
(\mathbf{W}_V \otimes \mathbf{I}_V) \cdot \mathbf{P}_V^V &= \left(\mathbf{I}_2 \otimes \left((\mathbf{W}_{V/2} \otimes \mathbf{I}_{V/2}) \cdot \mathbf{P}_{V/2}^{V/2}\right) \otimes \mathbf{I}_2\right)^{\left(\mathbf{P}_2^{V/2} \otimes \mathbf{I}_V\right)} \\
&\cdot \left(\mathbf{I}_{V/2} \otimes \mathbf{D}_{V/2}^2 \otimes \mathbf{I}_2\right) \cdot \left(\mathbf{I}_{V/2} \otimes \mathbf{W}_2 \otimes \mathbf{I}_V\right) \cdot \left(\mathbf{I}_{V/2} \otimes \mathbf{P}_V^2\right)
\end{aligned} \quad (4.55)$$

Formula (4.55) defines the radix-2 butterfly stages and the permutation stages recursively based on $(\mathbf{W}_{V/2} \otimes \mathbf{I}_{V/2}) \cdot \mathbf{P}_{V/2}^{V/2}$. Each step of the recursion adds one radix-2 FFT stage and one permutation stage on elements of pairs of vectors. The permutation defined by $\left(\mathbf{P}_2^{V/2} \otimes \mathbf{I}_V\right)$ performs a reordering of complete data vectors, which is later reversed.

Permutation stages based on butterfly permutations

The permutations in the above-described algorithms are based on stride permutations on pairs of vectors (see e.g. figures 4.4 and 4.5). The required stride permutations have two drawbacks concerning the implementation on the scalable SIMD processor architecture: Firstly, none of the implemented permutation networks directly supports stride permutation by specialized instructions. Hence, stride permutations have to be defined manually using permutation registers to store the stride permutation pattern. Furthermore, some stride permutations potentially cannot be realized on an inverse butterfly network in one permutation operation.

Secondly, stride permutations on pairs of vectors are relatively complex, as each output vector contains permuted elements of both input vectors. This has no impact on the realization on double-vector networks, yet multiple consecutive permutation operations are required on a single-vector network. An example for $\mathbf{P}_4^2$ and $V = 4$ is shown in figure 4.7. On a single-vector permutation network, $\mathbf{P}_4^2$ may be realized by a pair of masked butterfly permutations (masking is represented by light gray boxes) and a pair of concluding permutations on single vectors.

Masked butterfly permutations are well suited for an FFT implementation, as they are directly supported by specialized instructions and enable permuting values from two vectors efficiently. Each output vector of a masked butterfly permutation contains $V/2$ non-permuted elements from one input vector, which are preserved by masking, and $V/2$ permuted elements from the second input. Figure

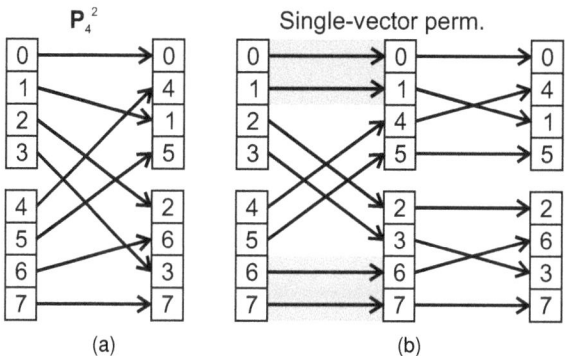

Figure 4.7: Stride permutation (a) based on $\mathbf{P}_4^2$ for $V=4$ and realization on a single-vector permutation network (b)

4.8 shows the possible masked butterfly permutations on pairs of vectors for a vector length of eight, as well as the corresponding representation by permutation matrix formulas.
Any masked butterfly permutation P_{Bfy}^B on pairs of vector, which permutes blocks of B elements ($B = 2^b < V$), can be expressed by permutation matrices based on stride permutations:

$$\mathbf{P}_{\text{Bfy}}^B = \left(\prod_{i=1}^{\log_2(V/B)-1} \left(\mathbf{I}_{2^{i-1}} \otimes \mathbf{P}_2^2 \otimes \mathbf{I}_{V/2^i} \right) \right) \cdot \left(\mathbf{P}_{V/B}^2 \otimes \mathbf{I}_B \right) \tag{4.56}$$

A vectorized FFT algorithm applies permutation operations on vector elements to perform the sorting of elements — for example, a bit-reversal for the pure radix-2 FFT — and for processing vector elements together during the radix-2 FFT stages. The latter is necessary, as every output element of the FFT is a weighted sum of all the input elements. A sequence of $\log_2(V)$ masked butterfly permutations with decreasing blocks sizes ($B \in [V/2, V/4, \ldots, 2, 1]$), interleaved with radix-2 FFT stages, suffices for the joint processing of vector elements during radix-2 stages. The sorting of FFT outputs can be realized by permutations of complete vectors if the FFT size is a multiple of the squared vector length. If the FFT size is not a multiple of the squared vector length, a final permutation stage is necessary for the sorting.

This is illustrated by the following example of an 8-point FFT for a vector length of four, based on formula (4.30). First, the permutation matrix $\mathbf{P}_4^2$ is expanded using formula (4.18). The resulting permutation $\mathbf{P}_2^2 \otimes \mathbf{I}_2$ is already a butterfly permutation; permutation $\mathbf{I}_2 \otimes \mathbf{P}_2^2$ can be moved to the next permutation stage on the left. Here, $\left(\mathbf{D}_2^2 \otimes \mathbf{I}_2 \right)^{(\mathbf{I}_2 \otimes \mathbf{P}_2^2)}$ results in a modified ordering of twiddle factors (based on formula (4.23)), but does not influence the vectorization.

$$\mathbf{W}_8 = (\mathbf{W}_2 \otimes \mathbf{I}_4) \cdot \left(\mathbf{P}_2^2 \otimes \mathbf{I}_2 \right) \cdot \left(\mathbf{D}_2^2 \otimes \mathbf{I}_2 \right) \cdot (\mathbf{W}_2 \otimes \mathbf{I}_4) \cdot \mathbf{P}_4^2 \cdot \mathbf{D}_4^2 \cdot (\mathbf{W}_2 \otimes \mathbf{I}_4)$$

Chapter 4 Radix-2 and mixed-radix FFTs for OFDM-A and SC-FDMA

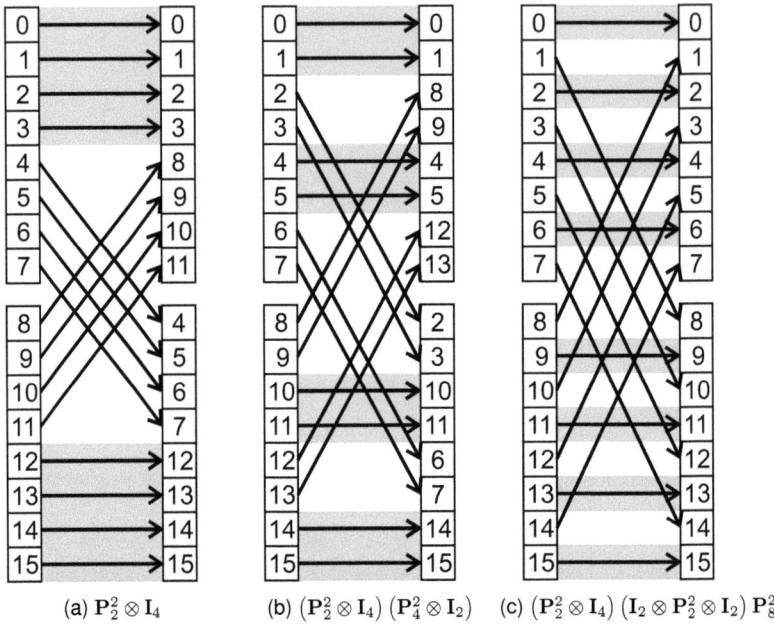

Figure 4.8: Masked butterfly permutations with different block sizes B for a vector length of eight: (a) $B = 4$, (b) $B = 2$, (c) $B = 1$

$$\begin{aligned}
&= (\mathbf{W}_2 \otimes \mathbf{I}_4) \cdot (\mathbf{P}_2^2 \otimes \mathbf{I}_2) \cdot (\mathbf{D}_2^2 \otimes \mathbf{I}_2) \\
&\quad \cdot (\mathbf{W}_2 \otimes \mathbf{I}_4) \cdot \underbrace{(\mathbf{I}_2 \otimes \mathbf{P}_2^2) \cdot (\mathbf{P}_2^2 \otimes \mathbf{I}_2)}_{\mathbf{P}_4^2} \cdot \mathbf{D}_4^2 \cdot (\mathbf{W}_2 \otimes \mathbf{I}_4)
\end{aligned} \tag{4.57}$$

$$\begin{aligned}
\mathbf{W}_8 &= (\mathbf{W}_2 \otimes \mathbf{I}_4) \cdot (\mathbf{P}_2^2 \otimes \mathbf{I}_2) \cdot (\mathbf{I}_2 \otimes \mathbf{P}_2^2) \cdot (\mathbf{D}_2^2 \otimes \mathbf{I}_2)^{(\mathbf{I}_2 \otimes \mathbf{P}_2^2)} \\
&\quad \cdot (\mathbf{W}_2 \otimes \mathbf{I}_4) \cdot (\mathbf{P}_2^2 \otimes \mathbf{I}_2) \cdot \mathbf{D}_4^2 \cdot (\mathbf{W}_2 \otimes \mathbf{I}_4)
\end{aligned} \tag{4.58}$$

Next, the formula is expanded by $(I_2 \otimes P_2^2) \cdot (I_2 \otimes P_2^2)$. The expression can be simplified by applying formula (4.18). Finally, permutations are reordered using formula (4.16).

$$\begin{aligned}
W_8 &= \underbrace{(I_2 \otimes P_2^2) \cdot (I_2 \otimes P_2^2)}_{I_8} \cdot (W_2 \otimes I_4) \cdot (P_2^2 \otimes I_2) \cdot (I_2 \otimes P_2^2) \cdot (D_2^2 \otimes I_2)^{(I_2 \otimes P_2^2)} \\
&\quad \cdot (W_2 \otimes I_4) \cdot (P_2^2 \otimes I_2) \cdot D_4^2 \cdot (W_2 \otimes I_4) \\
&= (I_2 \otimes P_2^2) \cdot (W_2 \otimes I_4) \cdot \underbrace{(I_2 \otimes P_2^2) \cdot (P_2^2 \otimes I_2)}_{P_4^2} \cdot (I_2 \otimes P_2^2) \cdot (D_2^2 \otimes I_2)^{(I_2 \otimes P_2^2)} \\
&\quad \cdot (W_2 \otimes I_4) \cdot (P_2^2 \otimes I_2) \cdot D_4^2 \cdot (W_2 \otimes I_4) \\
&= (I_2 \otimes P_2^2) \cdot (W_2 \otimes I_4) \cdot P_4^2 \cdot (I_2 \otimes P_2^2) \cdot (D_2^2 \otimes I_2)^{(I_2 \otimes P_2^2)} \\
&\quad \cdot (W_2 \otimes I_4) \cdot (P_2^2 \otimes I_2) \cdot D_4^2 \cdot (W_2 \otimes I_4) \\
&= \underbrace{(I_2 \otimes P_2^2)}_{\text{final perm.}} \cdot (W_2 \otimes I_4) \cdot \underbrace{(P_2^2 \otimes I_2) \cdot P_4^2}_{P_{\text{Bfy}}^1} \cdot (D_2^2 \otimes I_2)^{(I_2 \otimes P_2^2)} \\
&\quad \cdot (W_2 \otimes I_4) \cdot \underbrace{(P_2^2 \otimes I_2)}_{P_{\text{Bfy}}^2} \cdot D_4^2 \cdot (W_2 \otimes I_4)
\end{aligned} \quad (4.59)$$

The resulting FFT algorithm comprises two masked butterfly permutations and one final permutation stage that produces the correct ordering of output values. Figure 4.9 depicts the resulting signal flow graph, the signal flow graph for the same FFT size with stride permutations is shown in figure 4.3.

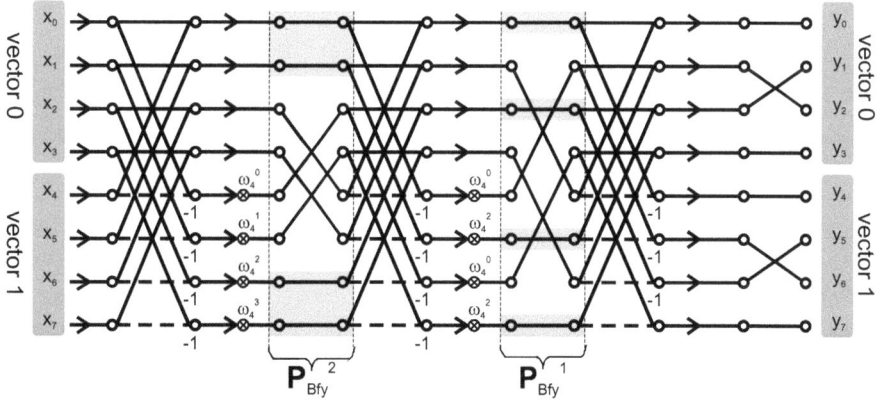

Figure 4.9: 8-point SIMD FFT algorithm with butterfly permutations

The implementations of radix-2 and mixed-radix FFT algorithms in the next section replace all stride permutations by butterfly permutations. For FFT sizes that are a multiple of the squared vector length, $\log_2(V)$ masked butterfly permutations are required. Short radix-2 FFTs ($2 \cdot V \leq N_{\text{DFT}} < V^2$) on processors with single-vector permutation networks require one additional permutation stage at the end of the FFT for sorting the outputs. On processors with double-vector permutation networks, the sorting can be realized by replacing the last butterfly permutation stage with a more elaborate permutation stage on pairs of vectors.

4.5 Radix-2 and mixed-radix FFT implementations based on LTE

LTE requires FFT sizes between 128 and 2048 for the IDFT at the transmitter (and the DFT at the receiver). The DFT-spreading in SC-FDMA is done by mixed-radix FFTs with lengths ranging from 12 to 1200. All required radix-2 FFTs have been implemented. Furthermore, all remaining short radix-2 FFTs that satisfy the constraint that the length of the FFT is at least twice the SIMD vector length have been implemented. The short radix-2 FFTs allow investigating the impact of an increased permutation complexity on the performance for different permutation networks. Only some mixed-radix FFTs have been implemented, because the constraint that the FFT size should be a multiple of the squared vector length prevents an implementation based on the proposed mixed-radix FFT algorithm for many FFT sizes. Here, the SIMD vector length V refers to the number of 32-bit elements in a vector, as each element of the FFT consists of a 16-bit real part and a 16-bit imaginary part. Table 4.2 summarizes the constraints for different vector lengths.

Table 4.2: Constraints for radix-2 and mixed-radix FFT sizes for different vector lengths

SIMD bit width	128	256	512	1024
Vector length [32-bit elements]	4	8	16	32
Minimum radix-2 FFT size	8	16	32	64
Minimum mixed-radix FFT size	$16 \cdot M$	$64 \cdot M$	$256 \cdot M$	$1024 \cdot M$

All mixed-radix FFTs required for SC-FDMA in LTE that satisfy the constraint for a SIMD bit width of 256 bit have been implemented on the 128-bit and 256-bit SIMD processors. A 768-point FFT algorithm has also been implemented on the 512-bit SIMD processors. Hence, the speedup, when the constraint is fulfilled can be measured. Furthermore, 384-point FFT algorithms for 512-bit and 1024-bit processors have been implemented, enabling to measure the performance for the case that the constraint is not fulfilled. Table 4.3 lists all implemented FFTs.

Before discussing the performance of the implemented FFTs in section 4.6, relevant information on the implementation is provided. First, the grouping of FFT stages to avoid memory access is explained. Next, the algorithms for the calculation of basic radix-2, radix-3, radix-5, and radix-6 butterfly stages are introduced. Afterwards, differences between the implementation of permutation operations on the various permutation networks are discussed. In the last part of this section, the implementation of short mixed-radix FFTs that cannot be realized by the mixed-radix FFT algorithm is explained.

Table 4.3: Implemented radix-2 and mixed-radix FFTs

FFT size	8-pt.	16-pt.	32-pt.	64-pt.	128-pt.	256-pt.	512-pt.	1024-pt.	2048-pt.	192-pt.	384-pt.	576-pt.	768-pt.	960-pt.	1152-pt.
128 bit	✓	✓	✓	✓	✓	✓	✓	✓	✓	✓	✓	✓	✓	✓	✓
256 bit		✓	✓	✓	✓	✓	✓	✓	✓	✓	✓	✓	✓	✓	✓
512 bit			✓	✓	✓	✓	✓	✓	✓		✓		✓		
1024 bit			✓	✓	✓	✓	✓	✓	✓						

4.5.1 Grouping of FFT stages

As the scalable SIMD processor architecture has a limited number of vector registers (16 general-purpose SIMD vector registers, see chapter 3.1.4), not all FFTs can be implemented without spilling data to memory between FFT stages. Short radix-2 FFTs that require at most eight data vectors[2] can be implemented in a single step, without spilling. The maximum FFT size for processing the FFT in a single step is $8 \cdot V$; hence, the three shortest implemented radix-2 FFTs can be implemented in one step for all vector lengths. The FFT sizes are listed in table 4.4.

Table 4.4: Short radix-2 FFTs that fit into the vector register file

SIMD width	128 bit	256 bit	512 bit	1024 bit
FFT sizes	8, 16, 32	16, 32, 64	32, 64, 128	64, 128, 256

Longer radix-2 and mixed-radix FFTs have to be split into groups of consecutive FFT stages, which process a subset of the complete DFT. The complete processing of the grouped FFT stages is realized by loops on the input data. The grouping of FFT stages is done to achieve a good ratio between *computational operations* on the VALU and VMAC and memory access operations on the VLSU. The computational operations cannot be avoided, as they are necessary for the FFT algorithm. Hence, the utilization of the VALU and the VMAC is a lower boundary for the runtime of a loop on a LIW processor architecture. Useful VALU and VMAC operations and memory access operations can be performed in parallel in one LIW operation. If the number of memory access operations is smaller than or equal to the number of useful operations on the VALU or VMAC, an overhead due to memory access can potentially be avoided by efficient LIW programming. If more memory access operations are required than computational operations on the VALU or VMAC, the runtime is determined by the number of memory access operations.

As each FFT stage operates on complete vectors and consecutive stages process different data values, the register demand increases with the number of consecutive DFT stages. Next to the registers for input data, further registers are required for twiddle factors and intermediate results. In

---
[2]The remaining vector registers are used for twiddle factor vectors and intermediate results.

particular, radix-3, radix-5, and radix-6 FFT stages require many data vectors for intermediate results (see section 4.5.2). Based on these restrictions, at most three consecutive radix-2 stages can be grouped together (eight input data vectors). Radix-5 and radix-6 stages cannot be efficiently grouped together with other FFT stages, due to the high register demand for intermediate values. Multiple radix-3 stages also cannot be grouped together, yet a radix-3 stage can potentially be combined with one or two radix-2 FFT stages (depending on the number of registers required for twiddle factors). In the majority of cases, consecutive radix-3 and radix-2 stages should be replaced by a radix-6 stage, which has a lower computational complexity (see section 4.5.2).

In case only two radix-2 FFT stages can be grouped together (i. e. all other stages already have been grouped together), the number of operations for loading and storing data is the same as the number of useful operations on the VALU and the VMAC (see section 3.1.4, table 3.6). In case further memory access is necessary for twiddle factors, the runtime is dominated by memory access. A single radix-2 FFT stage always requires more clock cycles on the VLSU than on the VALU or the VMAC.

Table 4.5 lists the decompositions of radix-2 and mixed-radix FFTs into groups of FFT stages for different SIMD widths. Groups of FFT stages, whose performance is degraded by memory access, are emphasized by using bold font and underlines (e. .g. **2**). In most cases, the decomposition into groups of FFT stages is the same for all SIMD vector lengths. The only exceptions are the 1024-point and the 384-point FFTs. The 384-point FFT requires a different grouping of FFT stages on 512-bit and 1024-bit SIMD processors than on processors with a smaller SIMD width, as a different FFT algorithm is used, because the constraint for the vectorization of the mixed-radix FFT is not satisfied (see section 4.5.4).

Table 4.5: Decomposition of long radix-2 and mixed-radix FFTs into groups of FFT stages in loops. The notation 2^x means that x radix-2 stages are grouped together.

SIMD bit width	128 bit	256 bit	512 bit	1024 bit
64-pt. FFT	$2^3, 2^3$	short FFT	short FFT	short FFT
128-pt. FFT	$2^3, \underline{\mathbf{2}}, 2^3$	$2^3, \underline{\mathbf{2}}, 2^3$	short FFT	short FFT
256-pt. FFT	$2^3, 2^2, 2^3$	$2^3, 2^2, 2^3$	$2^3, 2^2, 2^3$	short FFT
512-pt. FFT	$2^3, 2^3, 2^3$	$2^3, 2^3, 2^3$	$2^3, 2^3, 2^3$	$2^3, 2^3, 2^3$
1024-pt. FFT	$2^3, 2^3, \underline{\mathbf{2}}, 2^3$	$2^3, 2^3, \underline{\mathbf{2}}, 2^3$	$2^3, 2^3, \underline{\mathbf{2}}, 2^3$	$2^3, 2^2, 2^2, 2^3$
2048-pt. FFT	$2^3, 2^3, 2^2, 2^3$	$2^3, 2^3, 2^2, 2^3$	$2^3, 2^3, 2^2, 2^3$	$2^3, 2^3, 2^2, 2^3$
192-pt. FFT	$2^3, 3, 2^3$	$2^3, 3, 2^3$	—	—
384-pt. FFT	$2^3, 6, 2^3$	$2^3, 6, 2^3$	$2^3, 3 \cdot 2, 2^3$	$\underline{\mathbf{2^2}}, 2 \cdot 3 \cdot 2, \underline{\mathbf{2}}, 2^2$
576-pt. FFT	$2^3, \underline{\mathbf{3}}, 3, 2^3$	$2^3, 3, 3, 2^3$	—	—
768-pt. FFT	$2^3, 3, \underline{\mathbf{2^2}}, 2^3$	$2^3, 3, 2^2, 2^3$	$2^3, 3, 2^2, 2^3$	—
960-pt. FFT	$2^3, 5, 3, 2^3$	$2^3, 5, 3, 2^3$	—	—
1152-pt. FFT	$2^3, 6, 3, 2^3$	$2^3, 6, 3, 2^3$	—	—

The 128-bit, 256-bit, and 512-bit implementations of the 1024-point FFT comprise one separate radix-2 FFT stage, preceded by a group of three radix-2 FFT stages. The runtime of the separate radix-2 stages is determined by memory access, while the runtime of the group of three radix-2 stages is determined by useful computations. If these radix-2 stages are instead grouped in two pairs of radix-2 stages, the performance of both corresponding loops is determined by memory access for loading twiddle factors, leading to a slightly worse performance than with the proposed decomposition. On a 1024-bit SIMD processor, all required twiddle factor vectors can be stored in registers and no memory access operations during loops are needed for loading twiddle factors. Hence, a grouping of pairs of radix-2 FFT stages offers the best performance on a 1024-bit SIMD processor architecture.

Table 4.5 also shows that only few FFTs suffer from performance degradations due to memory access. Furthermore, increasing the SIMD width counteracts performance degradations due to memory access, as long as the vectorization constraints on the ratio between FFT size and SIMD width are still satisfied.

All implementations of FFTs that satisfy the constraints on the FFT size share common loops for groups of radix-2 FFT stages that can be reused for all FFT sizes — and in part also for all SIMD widths: The FFTs start and end with groups of three radix-2 FFT stages. The first group of radix-2 stages is the same for all FFT sizes and SIMD widths, only parameters, such as address offsets and twiddle factors, change. The last group of radix-2 stages performs the reordering of vector elements — or part of the reordering of vector elements — and can be used for all FFT implementations on the same SIMD processor architecture[3]. Radix-3, radix-5, and radix-6 FFT stages can be reused for different SIMD widths; they can also be reused for different FFT sizes as long as the necessary reordering of vectors is adjusted.

Memory requirements of the FFT algorithms

All short radix-2 FFTs, which can be realized by a single loop, can be performed *in place*, i.e. the input values are overwritten by the final output of the FFT. The memory requirements of longer radix-2 and mixed-radix FFTs depend on the grouping of FFT stages. An FFT can be implemented in place if the groups of FFT stages can perform the necessary reordering of data vectors.

FFTs with $N_{\text{DFT}} = V \cdot M \cdot V$ perform the reordering of data vectors during the M-point FFT.[4] If an M-point FFT fits into the register file, all necessary permutations of complete data vectors can be done in place. If M vectors do not fit into the register file, the FFT can only be computed in place if the permutation of data vectors can be split into a series of smaller permutation operations, which can be performed on the input or output of groups of FFT stages that fit into the register file. Otherwise, there is a small memory overhead for storing intermediate results during the sorting of vectors. The memory overhead can be avoided by inserting a separate sorting stage, at the cost of an increased runtime of the FFT, or by smartly overlapping memory read and write access for the same group of FFT stages on different input data, enabling to perform more complex permutations of complete vectors without memory access. The latter approach leads to an increased (doubled, tripled, or quadrupled) code size of the corresponding loop. Yet, the increase in code size is significantly lower than the decrease in data memory overhead.

---

[3] The implementation depends on the SIMD width and the selected permutation network.
[4] In general, the M-point FFT performs the reordering of blocks of X elements for any FFT with $N_{\text{DFT}} = X \cdot M \cdot X$.

4.5.2 Implementation of DFT stages

Next to efficiently decomposing the FFT into many small DFTs on entire vectors and grouping them together, the DFT stages also need to be implemented efficiently. The basic radix-2 DFT, can be directly implemented by the corresponding matrix operation, requiring one vector addition and one vector subtraction.

$$\mathbf{y} = \mathbf{W}_2 \cdot \mathbf{x} = \begin{bmatrix} 1 & 1 \\ 1 & -1 \end{bmatrix} \cdot \begin{bmatrix} x_0 \\ x_1 \end{bmatrix} \tag{4.60}$$

Radix-3, radix-5, and radix-6 DFTs can be realized using the corresponding DFT matrices, yet more efficient implementations exist that are briefly discussed below.

The radix-3 and radix-5 algorithms are based on Temperton [Tem83], with modifications to adjust for the representation of complex-valued data types on the scalable SIMD processor architecture. The radix-3 DFT $\mathbf{y} = \mathbf{W}_3 \cdot \mathbf{x}$ is implemented by the following operations:

$$
\begin{aligned}
t_0 &= x_1 + x_2 & t_1 &= x_0 - \frac{t_0}{2} \\
t_2 &= -i\frac{\sqrt{3}}{2} \cdot (x_1 - x_2) & y_0 &= x_0 + t_0 \\
y_1 &= t_1 + t_2 & y_2 &= t_1 - t_2
\end{aligned}
\tag{4.61}
$$

The radix-3 algorithm occupies the VALU for five clock cycles (two vector subtractions and three vector additions) and the VMAC for three clock cycles (one complex-valued multiplication, one real-valued MAC operation). Twiddle factor multiplications between the radix-3 stage and the following DFT stage can be merged with the radix-3 processing, increasing the number of operations on the VMAC by two complex-valued multiplications (four clock cycles).

The radix-5 DFT $\mathbf{y} = \mathbf{W}_5 \cdot \mathbf{x}$ is implemented as follows:

$$
\begin{aligned}
t_0 &= x_1 + x_4 & t_1 &= x_2 + x_3 \\
t_2 &= \sin(0.4\pi) \cdot (x_1 - x_4) & t_3 &= \sin(0.4\pi) \cdot (x_2 - x_3) \\
t_4 &= t_0 + t_1 \\
t_5 &= \frac{\sqrt{5}}{4}(t_0 - t_1) & t_6 &= x_0 - \frac{1}{4} \cdot t_4 \\
t_7 &= i \cdot \left(t_2 + \frac{\sin(0.2\pi)}{\sin(0.4\pi)} \cdot t_3\right) & t_8 &= i \cdot \left(t_3 - \frac{\sin(0.2\pi)}{\sin(0.4\pi)} \cdot t_2\right) \\
t_9 &= t_6 + t_5 & t_{10} &= t_6 - t_5 \\
y_0 &= x_0 + t_4 & y_1 &= t_9 - t_7 \\
y_2 &= t_{10} + t_8 & y_3 &= t_{10} - t_8 \\
y_4 &= t_9 + t_7
\end{aligned}
\tag{4.62}
$$

The radix-5 algorithm occupies the VALU for 13 clock cycles (seven vector additions and six vector subtractions) and the VMAC for five clock cycles (2 real-valued multiplications and three real-valued MAC operations). As in the radix-3 case, twiddle factor multiplications can be merged with the DFT stage, increasing the number of operations on the VMAC by four complex-valued multiplications (eight clock cycles).

The radix-6 DFT consists of two prime-factors; hence, the prime-factor DFT algorithm by Good [Goo58] can be applied, which avoids twiddle factor multiplications between the radix-3 and radix-2 stages by smartly reordering inputs and outputs:

$$\mathbf{y} = \mathbf{W}_6 \cdot \mathbf{x} \tag{4.63}$$

$$\Leftrightarrow \begin{bmatrix} y_0 \\ y_4 \\ y_2 \\ y_3 \\ y_1 \\ y_5 \end{bmatrix} = \begin{bmatrix} \mathbf{W}_3 & \mathbf{W}_3 \\ \mathbf{W}_3 & -\mathbf{W}_3 \end{bmatrix} \cdot \begin{bmatrix} x_0 \\ x_2 \\ x_4 \\ x_3 \\ x_5 \\ x_1 \end{bmatrix} \tag{4.64}$$

Due to the complexity of the permutations, the algorithm is not practical for bigger FFTs.

4.5.3 Implementation of permutation stages for different permutation networks

Permutations of vector elements are necessary for the masked butterfly permutation stages on pairs of vectors and for the final permutation stage of short radix-2 FFTs, which is more elaborate than the masked butterfly permutation stages.

Masked butterfly permutations on pairs of vectors

The implementation of masked butterfly permutations on pairs of vectors depends on the width of the permutation network. On a SIMD processor with a double-vector permutation network, a masked butterfly permutation can be realized by one masked permutation instruction that overwrites the values in the two input registers with permuted values based on the values of the vector mask.

On a SIMD processor with a single-vector permutation network, the same permutation requires two consecutive masked butterfly permutations and an additional move operation to copy one of the input operands. Figure 4.10 contains the assembly code for the equivalent of a masked butterfly permutation on a pair of vectors, implemented by permutations on a single-vector network. First, the second input vector is moved to a different vector register to preserve the values for the second permutation. The move operation can be performed on either the VALU, the VMAC, or the VPU. Next, the values of the first input are permuted; the second input register is overwritten with the permuted values based on a vector mask. In the third step, the same permutation is performed on the copied second input, the first input register is overwritten with the permuted values based on a second vector mask[5].

The masked butterfly permutation stages on pairs of vectors are interleaved with the processing of radix-2 FFT stages. On a SIMD processor with a single-vector permutation network, each permutation stage requires three operations (two permutations on the VPU, one vector move operation) per

---
[5]m2 is calculated by negating all elements of m1.

```
1: vmov_valu v3 v2
2: vbfy64 v2 v1 m1
3: vbfy64 v1 v3 m2
```

Figure 4.10: Assembly code for realizing the equivalent of a masked 64-bit butterfly permutation on a pair of vectors by permutations on a single-vector network

pair of vectors, while a radix-2 FFT stage requires two operations on the VALU and one complex-valued twiddle factor multiplication on the VMAC[6] per pair of vectors. As the permutation stage requires more operations than the radix-2 FFT stage, permutations cannot be totally hidden by LIW execution and the performance degrades compared to the performance of radix-2 FFT stages without permutations. The actual overhead depends on the number of vectors and the number of permutation stages.

On a SIMD processor with a double-vector permutation network, permutation stages require only one operation per pair of vectors. Hence, there is no overhead for permutations on a LIW SIMD processor, the permutations for one pair of vectors can be done in parallel to the radix-2 FFT butterfly on a different pair of vectors.

Final permutation stage of short radix-2 FFTs

Short radix-2 FFTs ($N_{DFT} < V^2$) require one permutation stage on pairs of vectors that is more complex than the usual masked butterfly permutation stages on pairs of vectors (see section 4.4.3). This permutation stage realizes the correct ordering of output values of the FFT. The complexity depends on the SIMD width and the FFT size. Different implementations of this permutation stage are required for the various permutation network types.

On a SIMD processor with a double-vector crossbar network, arbitrary permutations on pairs of vectors can be done in one permutation operation. Processors with single-vector networks can only perform masked butterfly permutations on pairs of vectors efficiently (as described in the previous paragraph); hence, the permutation stage is decomposed into a butterfly permutation stage on pairs of vectors (two permutation operations) and two final permutations on single vectors. On a processor architecture with a double-vector inverse butterfly network, many permutations on pairs of vectors can be realized in one operation. Yet, the last permutation stage of short radix-2 FFTs is too complex for an inverse butterfly network. Hence, the permutation is also implemented by two consecutive permutation stages. The added permutations can be interleaved with the processing of the last radix-2 FFT stage, reducing their overhead. Although, single-vector permutation networks and double-vector inverse butterfly network both require an additional permutation stage, the performance of the double-vector network is much better, as it requires only one operation per pair of vectors. Table 4.6 lists the overhead for additional permutations for different FFT sizes and SIMD widths. Some short radix-2 FFTs algorithms for 512-bit and 1024-bit SIMD processors require two additional permutation stages for inverse butterfly networks as the required permutation cannot be realized in one step. Yet, in this

---

[6]Except for the last FFT stage, which does not require a twiddle factor multiplication.

case, the performance of a double-vector inverse butterfly network is still better than the performance of a single-vector crossbar network.

Table 4.6: Overview of additional permutation stages for short radix-2 FFTs: Single-vector and double-vector inverse butterfly networks are denoted as Bfy1 and Bfy2, respectively, a single-vector crossbar network is denoted as Cross1.

SIMD width	FFT size	Additional permutation stages
128 bit	8	Bfy1, Bfy2, Cross1: + 1 perm.
256 bit	16	Bfy1, Bfy2, Cross1: + 1 perm.
	32	
512 bit	32	Bfy1, Bfy2: + 2 perm. Cross1: + 1 perm
	64	Bfy1, Bfy2, Cross1: + 1 perm.
	128	
1024 bit	64	Bfy1, Bfy2: + 2 perm. Cross1: + 1 perm
	128	
	256	Bfy1, Bfy2, Cross1: + 1 perm.
	512	

4.5.4 Short mixed-radix FFT implementation

Short mixed-radix FFTs with lengths that are not multiples of the squared SIMD vector length cannot be implemented with the proposed mixed-radix FFT algorithm. Examples for this case are the 384-point mixed-radix FFTs on 512-bit ($V = 16$) and 1024-bit ($V = 32$) SIMD processors.

384-point FFT on a 512-bit SIMD processor

On a 512-bit SIMD processor, the 384-point FFT may be factorized as $N_{\mathsf{DFT}} = V/2 \cdot 2 \cdot 3 \cdot V/2$. The following formula shows the decomposition of an FFT with $N_{\mathsf{DFT}} = V/2 \cdot 2 \cdot m \cdot V/2$ into three smaller transforms based on the basic FFT decomposition in equation (4.5):

$$\begin{aligned}\mathbf{W}_{V/2 \cdot m \cdot 2 \cdot V/2} &= \left(\mathbf{W}_{mV} \otimes \mathbf{I}_{V/2}\right) \cdot \mathbf{P}_{mV}^{V/2} \cdot \mathbf{D}_{mV}^{V/2} \cdot \left(\mathbf{W}_{V/2} \otimes \mathbf{I}_{mV}\right) \\ &= \left(\mathbf{W}_{V/2} \otimes \mathbf{I}_{mV}\right) \cdot \left(\mathbf{P}_{V/2}^{2m} \otimes \mathbf{I}_{V/2}\right) \cdot \left(\mathbf{D}_{V/2}^{2m} \otimes \mathbf{I}_{V/2}\right) \\ &\quad \cdot \left(\mathbf{W}_{2m} \otimes \mathbf{I}_{\frac{V^2}{4}}\right) \cdot \mathbf{P}_{mV}^{V/2} \cdot \mathbf{D}_{mV}^{V/2} \cdot \left(\mathbf{W}_{V/2} \otimes \mathbf{I}_{mV}\right)\end{aligned} \quad (4.65)$$

For $V \geq 4$, the FFT algorithm in formula (4.65) performs all FFT stages on complete vectors, yet the algorithm requires more complex permutations on vector elements, which depend on the factor m, compared to the mixed-radix FFT algorithm for FFT sizes that are a multiple of the squared vector length. As the performance analysis in section 4.6.2 will show, the increased permutation complexity of the 384-point FFT for a SIMD width of 512 bits leads to a slightly degraded performance.

384-point FFT on a 1024-bit SIMD processor

The algorithm in formula (4.65) does not work for a 384-point FFT on a 1024-bit SIMD processor, as the FFT size is too short compared to the vector length $V = 32$. The only way to achieve that all FFT stages operate on complete vectors is a technique, which in the following is denoted as the virtual reduction of the vector length. A virtual reduction of the vector length is done by additional masked butterfly operations on pairs of vectors, which perform a block interleaving of independent data vectors, i.e. data from different FFTs or from already processed FFT stages. Later on, the masked butterfly permutation has to be reversed. Each reduction stage virtually halves the vector length. An example for the block-interleaved processing of two FFTs is shown in figure 4.11.

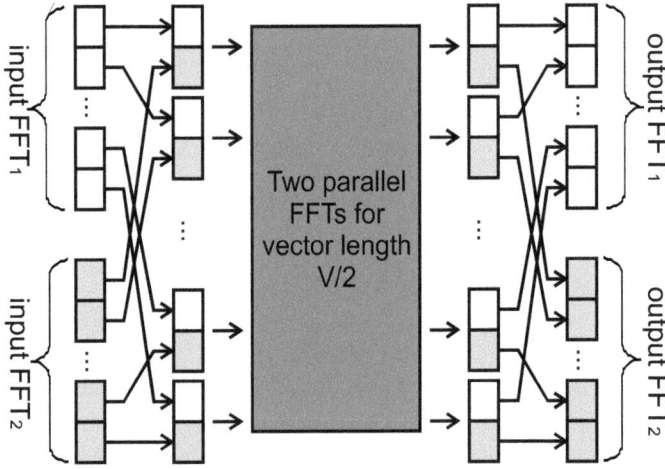

Figure 4.11: Merging of two FFTs for virtually reducing the vector length

The main drawback of this approach, besides requiring further permutations, is that *independent* data vectors, which are no longer processed together, have to be merged. During the mixed-radix FFT algorithm for $N_{DFT} = V \cdot M \cdot V$ and the radix-2 FFT algorithm for FFT sizes that are at least twice the SIMD width, masked butterfly permutation stages always operate on pairs of vectors that are processed together in the following radix-2 FFT stage. Therefore, permutation stages do not require additional data vectors, which occupy registers, and do not influence the grouping of permutation stages. If a virtual reduction of the vector length is done, pairs of independent data vectors have to be available in registers, which reduces the amount of registers available for grouping FFT stages by 50 percent: At most two consecutive radix-2 FFT stages can be grouped together if a virtual reduction is done (or reversed) in the current step. This leads to a significantly increased number of memory access dominated loops containing one or two consecutive FFT stages. In case of the 384-point FFT implementation, the permutations for the virtual reduction of the vector length can be hidden by LIW execution, a performance degradation (see section 4.6.2) solely occurs due to an increased number of memory access operations.

4.6 Performance analysis

The analysis of the performance of the implemented radix-2 and mixed-radix FFT algorithms on the scalable SIMD processor architecture is done in multiple steps. First, the achievable throughputs are reported and compared to the requirements of SISO and MIMO LTE channels. Next, the scalability of the FFT algorithms is analyzed by a discussion of speedup results. The parameters of the speedup measurements are adjusted based on the comparison to LTE requirements. Afterwards, radix-2, radix-3, radix-5, and radix-6 FFT loops are analyzed regarding LIW resource utilization and performance. In the last part of this section, the performance results are compared to FFT implementations on other SDR architectures in the literature.

4.6.1 Overview of throughput results

Table 4.7 lists the peak throughput of the FFT algorithms for SIMD processors clocked at 300 MHz. The throughput is measured in FFTs per second. The nomenclature of permutation networks is based on table 3.9 (chapter 3). The peak throughput is achieved if a processor is only performing FFT computations and the initialization of parameters can be neglected. For LTE-based systems, this assumption is not reasonable as a comparison to the required throughput shows.

LTE defines a short (seven OFDM symbols per slot) and a long CP (six OFDM symbols per slot) mode. Each slot has a duration of 0.5 ms. Hence, a throughput of $1.4 \cdot 10^4$ FFTs per second is required for a SISO system in short cyclic-prefix mode, in long cyclic-prefix mode the required throughput is $1.2 \cdot 10^4$ FFTs per second. For MIMO systems, the throughput requirement increases linearly with the number of antennas, as FFTs have to be computed for each antenna. Hence, the highest throughput requirement is $5.6 \cdot 10^4$ FFTs per second for a 4×4 MIMO system in short CP mode.

All FFT implementations in table 4.7 achieve better throughputs than required for LTE. Furthermore, the difference between the possible throughput on the SIMD processor and the required throughput can reach two orders of magnitude for short FFT lengths and/or long vector lengths. Hence, the SIMD processors might perform other useful computations in between the FFT processing for one slot. Therefore, throughput and speedup should be measured for a single slot. In this case, the overhead for initializing parameters cannot be neglected, especially for short FFTs. The worst-case scenario is the long CP mode for SISO transmission, which requires only six FFTs per slot. Therefore, this scenario has been selected for the speedup calculations in the following section. The achieved throughputs for this scenario are listed in table 4.8 for the sake of completeness.

4.6.2 Speedup results

Figure 4.12 contains speedup results for the implemented radix-2 and mixed-radix FFTs. The speedup is measured by normalizing throughput results to the throughput of a 128-bit SIMD proces-

Table 4.7: Peak throughput in FFTs per second without overhead for initialization

FFT size	128-bit SIMD: FFTs/s				256-bit SIMD: FFTs/s			
	Bfy1	Cross1	Bfy2	Cross2	Bfy1	Cross1	Bfy2	Cross2
8	$4.50 \cdot 10^7$	$4.50 \cdot 10^7$	$5.00 \cdot 10^7$	$5.00 \cdot 10^7$	$3.33 \cdot 10^7$	$3.33 \cdot 10^7$	$3.75 \cdot 10^7$	$3.75 \cdot 10^7$
16	$1.88 \cdot 10^7$	$1.88 \cdot 10^7$	$1.88 \cdot 10^7$	$1.88 \cdot 10^7$	$1.43 \cdot 10^7$	$1.43 \cdot 10^7$	$1.50 \cdot 10^7$	$1.50 \cdot 10^7$
32	$7.50 \cdot 10^6$	$7.50 \cdot 10^6$	$7.50 \cdot 10^6$	$7.50 \cdot 10^6$	$6.00 \cdot 10^6$	$6.00 \cdot 10^6$	$6.25 \cdot 10^6$	$6.25 \cdot 10^6$
64	$3.06 \cdot 10^6$	$3.06 \cdot 10^6$	$3.13 \cdot 10^6$	$3.13 \cdot 10^6$	$2.27 \cdot 10^6$	$2.27 \cdot 10^6$	$2.34 \cdot 10^6$	$2.34 \cdot 10^6$
128	$1.15 \cdot 10^6$	$1.15 \cdot 10^6$	$1.17 \cdot 10^6$	$1.17 \cdot 10^6$	$1.14 \cdot 10^6$	$1.14 \cdot 10^6$	$1.17 \cdot 10^6$	$1.17 \cdot 10^6$
256	$5.77 \cdot 10^5$	$5.77 \cdot 10^5$	$5.86 \cdot 10^5$	$5.86 \cdot 10^5$	$5.07 \cdot 10^5$	$5.07 \cdot 10^5$	$5.21 \cdot 10^5$	$5.21 \cdot 10^5$
512	$2.57 \cdot 10^5$	$2.57 \cdot 10^5$	$2.60 \cdot 10^5$	$2.60 \cdot 10^5$	$2.08 \cdot 10^5$	$2.08 \cdot 10^5$	$2.13 \cdot 10^5$	$2.13 \cdot 10^5$
1024	$1.05 \cdot 10^5$	$1.05 \cdot 10^5$	$1.07 \cdot 10^5$	$1.07 \cdot 10^5$	$1.04 \cdot 10^5$	$1.04 \cdot 10^5$	$1.07 \cdot 10^5$	$1.07 \cdot 10^5$
2048	$5.27 \cdot 10^4$	$5.27 \cdot 10^4$	$5.33 \cdot 10^4$	$5.33 \cdot 10^4$	$1.46 \cdot 10^6$	$1.46 \cdot 10^6$	$1.50 \cdot 10^6$	$1.50 \cdot 10^6$
192	$7.39 \cdot 10^5$	$7.39 \cdot 10^5$	$7.50 \cdot 10^5$	$7.50 \cdot 10^5$	$6.76 \cdot 10^5$	$6.76 \cdot 10^5$	$6.94 \cdot 10^5$	$6.94 \cdot 10^5$
384	$3.42 \cdot 10^5$	$3.42 \cdot 10^5$	$3.47 \cdot 10^5$	$3.47 \cdot 10^5$	$3.82 \cdot 10^5$	$3.82 \cdot 10^5$	$3.91 \cdot 10^5$	$3.91 \cdot 10^5$
576	$1.87 \cdot 10^5$	$1.87 \cdot 10^5$	$1.89 \cdot 10^5$	$1.89 \cdot 10^5$	$2.95 \cdot 10^5$	$2.95 \cdot 10^5$	$3.02 \cdot 10^5$	$3.02 \cdot 10^5$
768	$1.47 \cdot 10^5$	$1.47 \cdot 10^5$	$1.49 \cdot 10^5$	$1.49 \cdot 10^5$	$2.12 \cdot 10^5$	$2.12 \cdot 10^5$	$2.17 \cdot 10^5$	$2.17 \cdot 10^5$
960	$1.07 \cdot 10^5$	$1.07 \cdot 10^5$	$1.08 \cdot 10^5$	$1.08 \cdot 10^5$	$1.80 \cdot 10^5$	$1.80 \cdot 10^5$	$1.84 \cdot 10^5$	$1.84 \cdot 10^5$
1152	$9.09 \cdot 10^4$	$9.09 \cdot 10^4$	$9.19 \cdot 10^4$	$9.19 \cdot 10^4$				

FFT size	512-bit SIMD: FFTs/s				1024-bit SIMD: FFTs/s			
	Bfy1	Cross1	Bfy2	Cross2	Bfy1	Cross1	Bfy2	Cross2
32	$2.25 \cdot 10^7$	$2.65 \cdot 10^7$	$3.00 \cdot 10^7$	$3.00 \cdot 10^7$	$1.91 \cdot 10^7$	$2.20 \cdot 10^7$	$2.50 \cdot 10^7$	$2.50 \cdot 10^7$
64	$1.18 \cdot 10^7$	$1.18 \cdot 10^7$	$1.25 \cdot 10^7$	$1.25 \cdot 10^7$	$8.96 \cdot 10^6$	$1.00 \cdot 10^7$	$1.07 \cdot 10^7$	$1.07 \cdot 10^7$
128	$4.76 \cdot 10^6$	$4.76 \cdot 10^6$	$5.36 \cdot 10^6$	$5.36 \cdot 10^6$	$4.17 \cdot 10^6$	$4.17 \cdot 10^6$	$4.62 \cdot 10^6$	$4.69 \cdot 10^6$
256	$2.27 \cdot 10^6$	$2.27 \cdot 10^6$	$2.34 \cdot 10^6$	$2.34 \cdot 10^6$	$1.76 \cdot 10^6$	$1.76 \cdot 10^6$	$2.08 \cdot 10^6$	$2.08 \cdot 10^6$
512	$1.01 \cdot 10^6$	$1.01 \cdot 10^6$	$1.04 \cdot 10^6$	$1.04 \cdot 10^6$	$8.72 \cdot 10^5$	$8.72 \cdot 10^5$	$9.38 \cdot 10^5$	$9.38 \cdot 10^5$
1024	$4.17 \cdot 10^5$	$4.17 \cdot 10^5$	$4.26 \cdot 10^5$	$4.26 \cdot 10^5$	$3.99 \cdot 10^5$	$3.99 \cdot 10^5$	$4.26 \cdot 10^5$	$4.26 \cdot 10^5$
2048	$2.08 \cdot 10^5$	$2.08 \cdot 10^5$	$2.13 \cdot 10^5$	$2.13 \cdot 10^5$	$2.21 \cdot 10^6$	$2.21 \cdot 10^6$	$2.33 \cdot 10^6$	$2.33 \cdot 10^6$
384	$1.28 \cdot 10^6$	$1.28 \cdot 10^6$	$1.34 \cdot 10^6$	$1.34 \cdot 10^6$				
768	$5.81 \cdot 10^5$	$5.81 \cdot 10^5$	$6.05 \cdot 10^5$	$6.05 \cdot 10^5$				

Table 4.8: Throughput in FFTs per second with overhead for initialization (long CP mode)

FFT size	128-bit SIMD: FFTs/s				256-bit SIMD: FFTs/s			
	Bfy1	Cross1	Bfy2	Cross2	Bfy1	Cross1	Bfy2	Cross2
8	$3.40 \cdot 10^7$	$3.40 \cdot 10^7$	$3.67 \cdot 10^7$	$3.83 \cdot 10^7$				
16	$1.68 \cdot 10^7$	$1.68 \cdot 10^7$	$1.68 \cdot 10^7$	$1.68 \cdot 10^7$				
32	$7.14 \cdot 10^6$	$7.14 \cdot 10^6$	$7.14 \cdot 10^6$	$7.14 \cdot 10^6$				
64	$2.96 \cdot 10^6$	$2.96 \cdot 10^6$	$3.02 \cdot 10^6$	$3.02 \cdot 10^6$				
128	$1.14 \cdot 10^6$	$1.14 \cdot 10^6$	$1.16 \cdot 10^6$	$1.16 \cdot 10^6$	$2.61 \cdot 10^7$	$2.61 \cdot 10^7$	$2.90 \cdot 10^7$	$3.00 \cdot 10^7$
256	$5.71 \cdot 10^5$	$5.71 \cdot 10^5$	$5.80 \cdot 10^5$	$5.80 \cdot 10^5$	$1.25 \cdot 10^7$	$1.25 \cdot 10^7$	$1.36 \cdot 10^7$	$1.36 \cdot 10^7$
512	$2.56 \cdot 10^5$	$2.56 \cdot 10^5$	$2.59 \cdot 10^5$	$2.59 \cdot 10^5$	$5.77 \cdot 10^6$	$5.77 \cdot 10^6$	$6.00 \cdot 10^6$	$6.00 \cdot 10^6$
1024	$1.05 \cdot 10^5$	$1.05 \cdot 10^5$	$1.06 \cdot 10^5$	$1.06 \cdot 10^5$	$2.21 \cdot 10^6$	$2.21 \cdot 10^6$	$2.28 \cdot 10^6$	$2.28 \cdot 10^6$
2048	$5.26 \cdot 10^4$	$5.26 \cdot 10^4$	$5.32 \cdot 10^4$	$5.32 \cdot 10^4$	$1.11 \cdot 10^6$	$1.11 \cdot 10^6$	$1.15 \cdot 10^6$	$1.15 \cdot 10^6$
192	$7.30 \cdot 10^5$	$7.30 \cdot 10^5$	$7.41 \cdot 10^5$	$7.41 \cdot 10^5$	$5.02 \cdot 10^5$	$5.02 \cdot 10^5$	$5.16 \cdot 10^5$	$5.16 \cdot 10^5$
384	$3.41 \cdot 10^5$	$3.41 \cdot 10^5$	$3.45 \cdot 10^5$	$3.45 \cdot 10^5$	$2.08 \cdot 10^5$	$2.08 \cdot 10^5$	$2.12 \cdot 10^5$	$2.12 \cdot 10^5$
576	$1.86 \cdot 10^5$	$1.86 \cdot 10^5$	$1.89 \cdot 10^5$	$1.89 \cdot 10^5$	$1.04 \cdot 10^5$	$1.04 \cdot 10^5$	$1.06 \cdot 10^5$	$1.06 \cdot 10^5$
768	$1.46 \cdot 10^5$	$1.46 \cdot 10^5$	$1.48 \cdot 10^5$	$1.48 \cdot 10^5$	$1.42 \cdot 10^6$	$1.42 \cdot 10^6$	$1.46 \cdot 10^6$	$1.46 \cdot 10^6$
960	$1.07 \cdot 10^5$	$1.07 \cdot 10^5$	$1.08 \cdot 10^5$	$1.08 \cdot 10^5$	$6.68 \cdot 10^5$	$6.68 \cdot 10^5$	$6.86 \cdot 10^5$	$6.86 \cdot 10^5$
1152	$9.07 \cdot 10^4$	$9.07 \cdot 10^4$	$9.17 \cdot 10^4$	$9.17 \cdot 10^4$	$3.79 \cdot 10^5$	$3.79 \cdot 10^5$	$3.88 \cdot 10^5$	$3.88 \cdot 10^5$
					$2.93 \cdot 10^5$	$2.93 \cdot 10^5$	$3.01 \cdot 10^5$	$3.01 \cdot 10^5$
					$2.11 \cdot 10^5$	$2.11 \cdot 10^5$	$2.16 \cdot 10^5$	$2.16 \cdot 10^5$
					$1.79 \cdot 10^5$	$1.79 \cdot 10^5$	$1.83 \cdot 10^5$	$1.83 \cdot 10^5$

FFT size	512-bit SIMD: FFTs/s				1024-bit SIMD: FFTs/s			
	Bfy1	Cross1	Bfy2	Cross2	Bfy1	Cross1	Bfy2	Cross2
32	$1.96 \cdot 10^7$	$2.17 \cdot 10^7$	$2.40 \cdot 10^7$	$2.54 \cdot 10^7$				
64	$1.05 \cdot 10^7$	$1.13 \cdot 10^7$	$1.15 \cdot 10^7$	$1.65 \cdot 10^7$				
128	$4.58 \cdot 10^6$	$4.58 \cdot 10^6$	$5.13 \cdot 10^6$	$5.17 \cdot 10^6$	$1.86 \cdot 10^7$	$2.05 \cdot 10^7$	$2.14 \cdot 10^7$	$1.00 \cdot 10^7$
256	$2.19 \cdot 10^6$	$2.19 \cdot 10^6$	$2.26 \cdot 10^6$	$2.26 \cdot 10^6$	$8.22 \cdot 10^6$	$9.09 \cdot 10^6$	$9.84 \cdot 10^6$	$4.52 \cdot 10^6$
512	$9.95 \cdot 10^5$	$9.95 \cdot 10^5$	$1.02 \cdot 10^6$	$1.02 \cdot 10^6$	$4.02 \cdot 10^6$	$4.02 \cdot 10^6$	$4.43 \cdot 10^6$	$2.01 \cdot 10^6$
1024	$4.13 \cdot 10^5$	$4.13 \cdot 10^5$	$4.23 \cdot 10^5$	$4.23 \cdot 10^5$	$1.71 \cdot 10^6$	$1.71 \cdot 10^6$	$2.01 \cdot 10^6$	$9.18 \cdot 10^5$
2048	$2.07 \cdot 10^5$	$2.07 \cdot 10^5$	$2.12 \cdot 10^5$	$2.12 \cdot 10^5$	$8.55 \cdot 10^5$	$8.55 \cdot 10^5$	$9.18 \cdot 10^5$	$4.22 \cdot 10^5$
384	$1.25 \cdot 10^6$	$1.25 \cdot 10^6$	$1.31 \cdot 10^6$	$1.31 \cdot 10^6$	$3.95 \cdot 10^5$	$3.95 \cdot 10^5$	$4.22 \cdot 10^5$	$2.22 \cdot 10^6$
768	$5.74 \cdot 10^5$	$5.74 \cdot 10^5$	$5.97 \cdot 10^5$	$5.97 \cdot 10^5$	$2.11 \cdot 10^6$	$2.11 \cdot 10^6$	$2.22 \cdot 10^6$	

sor with a single-vector inverse butterfly permutation network. Figure 4.12a contains speedup results based on the peak throughput measurement (without the overhead for initialization) and figure 4.12b shows the speedup based on performance results for one slot (long CP, SISO transmission). FFT sizes are displayed on the abscissa. Radix-2 FFTs are displayed on the left-hand side and mixed-radix FFTs on the right-hand side. Both diagrams show four groups of curves for 128-bit (speedup approximately one), 256-bit (speedup approximately two), 512-bit (speedup approximately four) and 1024-bit (speedup approximately eight) SIMD processors with different permutation networks.

Differences between figures 4.12a and 4.12b are mainly visible for short FFT sizes (e. g. the 64-point FFTs for 1024-point SIMD processors). The initialization overhead does not scale with the SIMD width; hence, the speedups for wider SIMD widths are reduced. The overhead for the initialization of parameters is insignificant compared to the runtime of FFT loops for longer FFTs.

If the constraints for the efficient vectorization of radix-2 and mixed-radix FFTs are satisfied, linear or close to linear speedup can be achieved. Deviations occur for small radix-2 FFTs, which require additional permutation stages (see section 4.5.3, table 4.6), and the 128-point and 1024-point FFTs. The last permutation stage of short radix-2 FFTs has an increased complexity and is decomposed into different permutation operations on the various permutation networks. On a double-vector crossbar network, no additional permutation is necessary and the highest throughput is achieved. The double-vector inverse butterfly network outperforms the single-vector networks and the single-vector crossbar network outperforms the single-vector inverse butterfly network. The impact of increasing the network width is more significant than the impact of changing from inverse butterfly to crossbar network.

Better than linear speedup is achieved for the 128-bit FFT on 512-bit and 1024-bit SIMD processors, because of a different grouping of FFT stages (see section 4.5.1). On processors with a SIMD width less than 512 bits, the 128-point FFT contains one loop that comprises just a single radix-2 FFT stage. The runtime of this loop on a LIW processor is determined by the number of memory access operations (two clock cycles per vectors), and not by the number of computational operations (one clock cycle per vector). On processors with a SIMD width of at least 512 bits, all seven FFT stages can be processed in one loop without any overhead for memory access. The 1024-point FFT implementations on 1024-bit SIMD processors also achieve better than linear speedups, because FFT stages can be grouped together more efficiently than on smaller SIMD processors.

The speedup results for the 384-point FFT on 512-bit and 1024-bit SIMD processors demonstrate the case of short mixed-radix FFTs, which do not satisfy the constraint that the FFT size should be a multiple of the squared SIMD vector length. The speedup on the 512-bit SIMD processors is only slightly worse than linear speedup, as the increased complexity can be efficiently compensated by LIW execution. On 1024-bit SIMD processors, the drop-off from linear speedup is significant, as permutations for the virtual reduction of the vector length prevent an efficient grouping of FFT stages, leading to an overhead for memory access operations.

4.6.3 Resource utilization and performance of FFT loops

Table 4.9 describes the utilization of LIW processing in radix-2, radix-3, radix-5, and radix-6 FFT loops. Loops whose performance is determined by memory access have been omitted, as an anal-

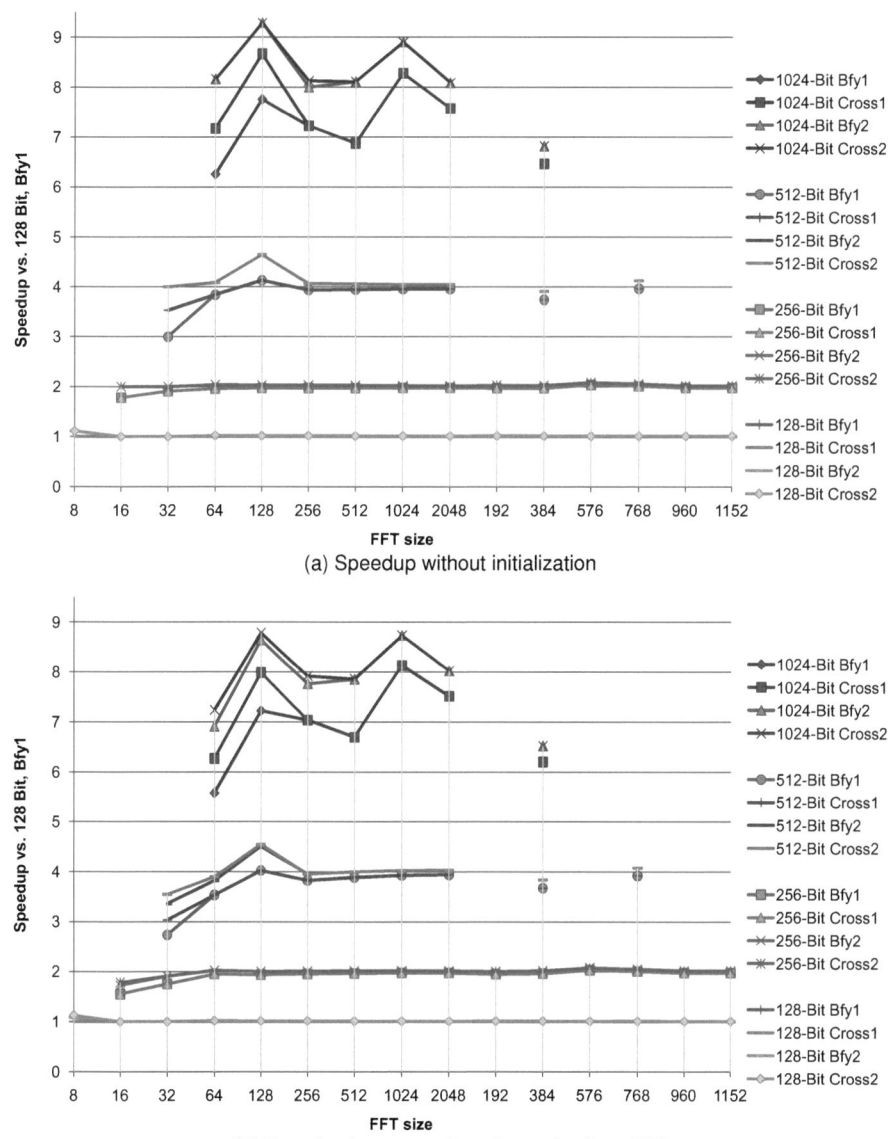

Figure 4.12: Speedup for radix-2 and mixed-radix FFTs on different SIMD processors measured versus a 128-bit SIMD processor with a single-vector butterfly network

ysis of resource utilization and throughput is mood in this case. The average number of operations per LIW instruction ($N_{par.\ \varnothing}$), depends on SIMD operations and scalar operations for loop dependent pointer update calculations. The pointer update calculations vary for different FFTs, leading to different values for $N_{par.\ \varnothing}$. The range of values has been stated in this case.

Table 4.9: Overview of the LIW performance of FFT loops: $N_{par.\ \varnothing}$ denotes the average number of parallel operations per instruction, the resource utilization of the VMAC and the VALU is denoted as R_{VMAC} and R_{VALU} respectively. The throughput is measured in clock cycles per output vector.

Description	SIMD proc.	$N_{par.\ \varnothing}$	R_{VMAC}	R_{VALU}	Cycles/ vector
3 radix-2 stages	all	2.92	100 %	100 %	3.00
2 radix-2 stages	all	2.56–3.3	100 %	100 %	2.00
3 radix-2 stages with perm.	128 bit, single-vector	3.12	64 %	96 %	3.13
	256 – 1024 bit, single-vector	3.5	61.5 %	92.3 %	3.25
	double-vector	2.75–3	66.7 %	100 %	3.00
Radix-3 stage	all	2.86–3.29	100 %	71.4 %	2.33
Radix-5 stage	all	2.82–3.16	87.5 %	81.3 %	3.20
Radix-6 stage	all	2.39–2.67	88.9 %	88.9 %	3.00

Radix-2 FFT loops without vector permutations achieve a throughput of one vector per clock cycle and radix-2 FFT stage. Both the VALU and the VMAC are utilized in each clock cycle. Radix-2 FFT loops with vector permutations achieve slightly lower throughputs on SIMD processors with single-vector permutation networks: The overhead for vector permutations cannot be completely mitigated by LIW execution (see section 4.5.3), which also leads to decreased resource utilization values, while the average number of parallel operations increases compared to a loop without permutations. The utilization of the VMAC is also reduced, because the last radix-2 FFT stage does not require twiddle factor multiplications. On a SIMD processor with a double-vector permutation network, permutations can be realized with fewer operations — and mitigated by LIW execution. The throughput is the same as in loops without permutations; the VALU is utilized all the time. Therefore, the runtime of arbitrary long radix-2 FFTs on SIMD processors with double-vector networks can be directly computed from the number of required operations on the VALU as one clock cycle per vector and radix-2 FFT stage. Radix-3, radix-5, and radix-6 FFT stages have an increased complexity compared to radix-2 FFT stages (see section 4.5.2). Therefore, the throughput is reduced compared to the radix-2 loops. High values for R_{VALU} and R_{VMAC} are achieved for all loops, indicating an efficient implementation of the FFT stages.

4.6.4 Comparison to other SDR FFT implementations

Table 4.10 compares the achieved performance on the proposed SIMD processor architecture to other SDR implementations on SIMD-based processors (EVP, SBX) and optimized FFT processors. The throughput of FFT implementations is measured in FFTs per second. All throughput measure-

Chapter 4 Radix-2 and mixed-radix FFTs for OFDM-A and SC-FDMA

ments are peak throughput measurements. The comparison is based on the proposed 256-bit SIMD processor with a single-vector inverse butterfly permutation network[7], as both referenced SIMD processors also support 256-bit SIMD operations.

Table 4.10: Comparison of SDR implementations of radix-2 and mixed-radix FFTs

Processor	256-Bit SIMD Bfy1	EVP [WBAHS08a]	SBX core [Beh09]	TTA ASP [PT09]	PFFT-M [AG09]
64-pt. FFTs/s	$6.00 \cdot 10^6$	$6.00 \cdot 10^6$	$5.35 \cdot 10^6$	$1.20 \cdot 10^6$	—
256-pt. FFTs/s	$1.14 \cdot 10^6$	$1.07 \cdot 10^6$	$1.34 \cdot 10^6$	—	$1.17 \cdot 10^6$
512-pt. FFTs/s	$5.07 \cdot 10^5$	$4.81 \cdot 10^5$	—	—	—
1024-pt. FFTs/s	$2.08 \cdot 10^5$	$2.11 \cdot 10^5$	$2.72 \cdot 10^5$	$4.84 \cdot 10^4$	$2.93 \cdot 10^5$
2048-pt. FFTs/s	$1.04 \cdot 10^5$	$9.77 \cdot 10^4$	$1.26 \cdot 10^5$	—	—
192-pt. FFTs/s	$1.46 \cdot 10^6$	$1.39 \cdot 10^6$	$\approx 8 \cdot 10^5$	N/A	—
384-pt. FFTs/s	$6.76 \cdot 10^5$	$6.15 \cdot 10^5$	$\approx 4 \cdot 10^5$	N/A	—
576-pt. FFTs/s	$3.82 \cdot 10^5$	$3.55 \cdot 10^5$	$\approx 2.7 \cdot 10^5$	N/A	—
768-pt. FFTs/s	$2.95 \cdot 10^5$	$2.71 \cdot 10^5$	$\approx 2.5 \cdot 10^5$	N/A	—
960-pt. FFTs/s	$2.11 \cdot 10^5$	$1.85 \cdot 10^5$	$\approx 1.2 \cdot 10^5$	N/A	—
1152-pt. FFTs/s	$1.80 \cdot 10^5$	$1.69 \cdot 10^5$	$\approx 1.4 \cdot 10^5$	N/A	—
Frequency	300 MHz	300 MHz	600 MHz	250 MHz	300 MHz
Power	74.7 mW + memories	$\approx$150 mW [SVPG+10]	$\approx$120 mW [San09]	60.4 mW	—
Technology	90 nm	45 nm	65 nm	130 nm	—
Area	0.83 mm^2 + memories	≈ 3 mm^2	—	280 kgates	—

The implementations on the EVP (see section 2.3.1) are based on the radix-2 and mixed-radix FFT algorithms in sections 4.4.1 and 4.4.2 [WBAHS08a, WS09a]. The EVP supports vector move operations on any vector unit; hence, the performance of the masked butterfly permutation stages on pairs of vectors can potentially be better than on the proposed scalable SIMD processor architecture. Yet, the throughput on the EVP is lower than the throughput of the 256-bit SIMD processor for almost all FFT sizes (exceptions are the 64-point and 1024-point FFTs). The performance on the EVP is worse due to the programming approach: The EVP is programmed in EVP-C, a programming language based on C-code for scalar operations and control flow and intrinsic vector operations for SIMD operations. The vectorization of algorithms is done by the programmer, yet the assignment of variables to registers and the scheduling of LIW instructions is done by a compiler. The compiler-generated code is usually less efficient than hand-coded assembly code.

The Sandblaster SB3500 processor architecture (see section 2.3.2) comprises three SBX processor cores. Each SBX processor core operates on 256-bit SIMD vectors and supports LIW execution. The performance results for one SBX core [Beh09] in table 4.10 are based on the maximum clock frequency of 600 MHz, which corresponds to four hardware threads at 150 MHz. Hence, the SBX core can perform twice as many MAC or multiply operations per second as the EVP and the proposed 256-

---

[7]Area and power consumption figures exclude memories.

bit SIMD processor at 300 MHz. However, the higher theoretical peak performance does not lead to significantly better throughput results, possibly due to a less efficient SIMD implementation of FFT algorithms. The throughput for radix-2 FFTs (except for the 64-point FFT) is between 17 and 30 percent higher than the throughput on the 256-bit SIMD processor. The mixed-radix FFT throughput is significantly worse than on the 256-bit SIMD processor.

The next processor architecture in table 4.10 is an application-specific processor (ASP) [PT09] based on the transport-triggered architecture (TTA). In a TTA, transfers of values are explicitly programmed in instructions and processing units are triggered by writing data to their input ports. Assuming multiple processing units and parallel connections, this approach achieves ILP. The proposed ASP supports power of two FFTs and contains complex-valued adder and multiplier units optimized for a radix-4 FFT algorithm. For the reported FFT sizes, the ASP achieves approximately one-fifth of the throughput of the 256-bit SIMD processor.

The PFFT-M processor [AG09] is an IP processor core for pipelined mixed-radix FFTs, specially designed for SDR applications, such as SC-FDMA in LTE. Radix-2, radix-3, radix-4, and radix-5 FFT stages are supported. The reported throughputs for 256-point and 1024-point FFTs are slightly better than the throughput results for the 256-bit SIMD processor.

4.7 Conclusion

In the previous sections, the development, the implementation, and the performance of radix-2 and mixed-radix FFT algorithms for SIMD processors have been discussed. The radix-2 FFT algorithm enables a vectorization with minimal overhead for vector element permutations for FFT sizes that are at least twice the vector length. The mixed-radix FFT algorithm requires the FFT size to be a multiple of the squared SIMD width.

The performance results show that support for LIW execution efficiently mitigates the overhead for vector permutation operations. In the majority of cases, the throughput is determined by the computational complexity of FFT stages and not by overhead due to the vectorization. Performance degradations occur due to FFT loops that are dominated by memory access.

The achieved throughput performance is competitive to dedicated FFT processors. The mixed-radix FFT implementations, based on the proposed algorithm, also outperform mixed-radix FFT implementations on the multi-threaded SIMD SBX processor.

The main drawback of the proposed mixed-radix FFT algorithm is that many of the FFTs for SC-FDMA cannot be vectorized using this algorithm on wide SIMD processors. Hence, the throughput for these FFTs will be reduced. However, the required throughput for LTE (SISO and MIMO transmission) can still be attained. Assuming that all FFT sizes, which are required for SC-FDMA in LTE, should be supported, enough resources (processing time on a SIMD processor) for achieving the required throughput for all FFTs should be allocated to the FFT processing. Short or efficiently vectorized FFTs will not require the full time slot, allowing either to perform other useful computations or to save power, e. g. by reducing the clock frequency and the supply voltage.

Chapter 5

Sphere decoding for MIMO detection

In the following, algorithms for symbol detection in spatial multiplexing multiple-input, multiple-output (MIMO) systems are explained. The discussion focuses on the fixed-complexity sphere decoder (FSD), which has been implemented on the scalable SIMD processor architecture.

Firstly, the MIMO system model and the sphere decoding principle are explained. In section 5.2, two modified sphere search algorithms with fixed-complexity are presented. Sections 5.3 and 5.4 focus on the FSD algorithm and its implementation for MIMO-OFDM on the scalable SIMD processor architecture. The next section contains an analysis of the achievable performance on the SIMD processor architecture and a comparison to other SDR and hardware implementations. Conclusions are drawn in section 5.6.

5.1 MIMO system model

Figure 5.1 shows a general MIMO system model. At the transmitter side, a signal or multiple signals are transmitted on n_T transmit antennas. The receiver receives n_R signals on different receive antennas. MIMO detection describes the task of reconstructing the transmitted signal or signals from the received signals.

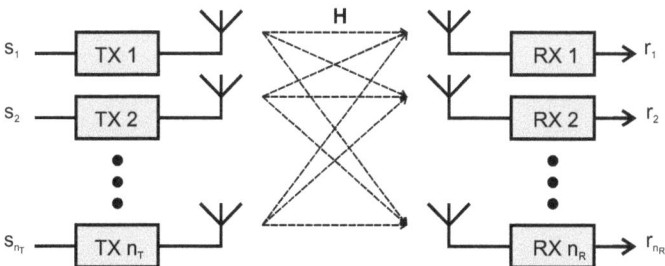

Figure 5.1: Channel model for a MIMO system with n_T transmit and n_R receive antennas

In principle, three different MIMO techniques exploit multiple antennas to improve signal quality or spectral efficiency [BÖ6, SBM+04]: *Diversity coding* aims at improved link reliability. One signal is transmitted over multiple antennas through — in the ideal case — independently fading channels.

The receiver compensates the fading by combining the received signals. An example for diversity coding is space-time block coding based on the Alamouti scheme [Ala98]. *Precoding* is a generalized beam-forming approach that requires knowledge of the MIMO channel: At the transmitter side, the signals that are emitted on the different antennas are weighted independently to maximize the signal power at the receiver side. This approach leads to an improved average signal-to-noise ratio (SNR) at the receiver. In *spatial multiplexing* MIMO systems, independent data streams are transmitted on the different antennas. The remainder of this section focuses on spatial multiplexing, as the obtainable capacity gain [FG98, Tel99] is necessary for high data rates in current and future wireless communication systems.

Assuming a *flat fading* channel[1], a spatial multiplexing MIMO system can be modeled by equation (5.1). Here, **H** is the channel matrix with i.i.d. complex-valued Gaussian elements, **r** and **s** are the received and transmitted signal vectors, respectively, and **n** is an additive white Gaussian noise vector.

$$r_j = \sum_{i=1}^{n_T} h_{i,j} \cdot s_i + n_j \qquad j \in [1, n_R]$$

$$\mathbf{r} = \mathbf{H} \cdot \mathbf{s} + \mathbf{n} \tag{5.1}$$

Spatial multiplexing offers a significant gain in channel capacity compared to a single-input, single-output (SISO) channel [FG98, Tel99]. For a SISO channel, the channel capacity C_{SISO} increases by 1 bps/Hz, if the SNR increases by 3 dB (see equation (5.2)). The capacity of a flat fading MIMO channel with perfect channel knowledge at the receiver (and no channel knowledge at the transmitter) is instead given by equation (5.3).

$$C_{SISO} = \log_2 \left(1 + SNR \cdot |H|^2\right) \tag{5.2}$$

$$C_{MIMO} = \log_2 \left(\det \left| \mathbf{I}_{n_R} + \frac{SNR}{n_T} \cdot \mathbf{H}\mathbf{H}^H \right| \right) \tag{5.3}$$

In principle, the channel capacity is determined by the minimum number of transmit and receive antennas $\min(n_T, n_R)$ [Bau01] with the SISO channel capacity as the lower boundary for $n_T = n_R = 1$. LTE [Tec06] supports spatial multiplexing transmission with $n_T = n_R = 2$ (2 × 2 MIMO) and optionally $n_T = n_R = 4$ (4 × 4 MIMO), WiMAX supports up to 4 × 4 MIMO [IEE09b]. In the following, the number of receive antennas and the number of transmit antennas are assumed to be the same. The implementation results in sections 5.4 and 5.5 are based on a 4 × 4 MIMO system.

---

[1]On a flat fading channel, all frequency components of the signal undergo the same magnitude of fading. OFDM can be used to decompose a frequency-selective channel into multiple flat fading channels [Bö6].

Chapter 5 Sphere decoding for MIMO detection

5.1.1 Maximum likelihood detection

The maximum likelihood (ML) detector for spatial multiplexing, determines the most likely transmitted signal vector $\hat{\mathbf{s}}_{ML}$ by minimizing the squared Euclidean distance of possible symbols transmitted over the channel to the received signal vector:

$$\hat{\mathbf{s}}_{ML} = \arg \min_{\mathbf{s} \in M^{n_T}} ||\mathbf{r} - \mathbf{H} \cdot \mathbf{s}||_2^2 \qquad (5.4)$$

Here, M denotes the modulation size, e.g. 16 for 16-QAM or four for QPSK. The ML detector achieves the optimum symbol detection results. Yet, in its direct implementation, the detector requires testing all valid symbols for each transmit antenna leading to a complexity of $\mathcal{O}(M^{n_T})$.

5.1.2 Sphere decoding

The sphere decoder (SD) calculates the ML solution with reduced computational complexity. Sphere decoding has been originally invented by Pohst [Poh81, FP85] and has its origin in finding the shortest vector in a lattice $\Lambda(\mathbf{H})$ [SE94, VB99, HtB03], defined by $\Lambda(\mathbf{H}) = \mathbf{H} \cdot \mathbf{s}$. The sphere decoder limits the minimum search in equation (5.4) to solutions within a sphere of radius d:

$$\hat{\mathbf{s}}_{SD} = \arg \left(\min_{\mathbf{s} \in M^{n_T}} ||\mathbf{r} - \mathbf{H} \cdot \mathbf{s}||_2^2 \leq d^2 \right) \qquad (5.5)$$

The SD algorithm consists of two steps: a transformation of the channel matrix into upper-triangular form and the sphere search.

The transformation of the channel matrix into an upper-triangular matrix can for example be done by a Cholesky factorization or a QR-decomposition. The QR-decomposition of a complex-valued channel matrix $\mathbf{H}$ (see equation (5.7)) decomposes the matrix into a *unitary* matrix $\mathbf{Q}$ and an upper-triangular matrix $\mathbf{R}$ [GVL96]. A unitary matrix is a complex-valued matrix that satisfies the condition:

$$\mathbf{Q}^H \cdot \mathbf{Q} = \mathbf{I}_{n_T} \qquad (5.6)$$

$$\mathbf{H} = \mathbf{Q} \cdot \mathbf{R} \qquad (5.7)$$

A QR-decomposition can for example be computed by successive Givens rotations, with each Givens rotation zeroing one channel matrix element [GVL96].
Using equation (5.7), the calculation of symbol vectors in a sphere can be rewritten as:

$$\begin{aligned} d^2 &\geq ||\mathbf{r} - \mathbf{Q} \cdot \mathbf{R} \cdot \mathbf{s}||_2^2 \\ \Leftrightarrow d^2 &\geq ||\mathbf{Q}^H \cdot \mathbf{r} - \mathbf{Q}^H \cdot \mathbf{Q} \cdot \mathbf{R} \cdot \mathbf{s}||_2^2 \\ \Leftrightarrow d^2 &\geq ||\mathbf{Q}^H \cdot \mathbf{r} - \mathbf{R} \cdot \mathbf{s}||_2^2 \\ \Leftrightarrow d^2 &\geq ||\mathbf{y} - \mathbf{R} \cdot \mathbf{s}||_2^2 \\ \text{with } \mathbf{y} &= \mathbf{Q}^H \cdot \mathbf{r} \end{aligned} \qquad (5.8)$$

The squared Euclidean distance in formula (5.8) can be calculated by accumulating partial (squared) Euclidean distances (PEDs) $d_i(\mathbf{s})$ defined by the following equation:

$$d_i(\mathbf{s}) = d_{i+1}(\mathbf{s}) + \left\| y_i - \sum_{j=i}^{n_T} R_{i,j} \cdot s_j \right\|_2^2 \qquad (5.9)$$
$$= d_{i+1}(\mathbf{s}) + e_i(\mathbf{s})$$

with $0 = d_{n_T+1}(\mathbf{s})$

The total squared Euclidean distance is given by $d_1(\mathbf{s})$. At each decoding level, the PED is incremented by $e_i(\mathbf{s})$. The sphere decoder requires the squared Euclidean distance to lie within a sphere defined by d^2. Hence, also the PED increments at level i must lie within the same sphere.

$$d^2 \geq d_1(\mathbf{s})$$
$$\Leftrightarrow d^2 \geq \sum_{i=1}^{n_T} e_i(\mathbf{s}) \qquad (5.10)$$
$$\Rightarrow d^2 \geq e_i(\mathbf{s}) \qquad \text{for } i \in [1, n_T]$$

Due to the upper-triangular structure of $\mathbf{R}$, a partial Euclidean distance at the i-th level depends only on transmitted symbols on antennas $i, i+1, \ldots, n_T$. This enables to perform the search for the most likely transmitted symbol vector by a tree search, with each level of the tree corresponding to one antenna symbol. The tree nodes represent PEDs, the edges between nodes from different tree levels represent the PED increments $e_i(\mathbf{s})$. For M-QAM, there are M branches at each level of the tree.

An example for sphere decoding in a 4×4 MIMO system with QPSK modulation is shown in figure 5.2. The PED increments at each decoding level are:

$$e_4(\mathbf{s}) = \|y_4 - R_{4,4} \cdot s_4\|_2^2 \qquad (5.11)$$
$$e_3(\mathbf{s}) = \|y_3 - R_{3,4} \cdot s_4 - R_{3,3} \cdot s_3\|_2^2 \qquad (5.12)$$
$$e_2(\mathbf{s}) = \|y_2 - R_{2,4} \cdot s_4 - R_{2,3} \cdot s_3 - R_{2,2} \cdot s_2\|_2^2 \qquad (5.13)$$
$$e_1(\mathbf{s}) = \|y_1 - R_{1,4} \cdot s_4 - R_{1,3} \cdot s_3 - R_{1,2} \cdot s_2 - R_{1,1} \cdot s_1\|_2^2 \qquad (5.14)$$

The SD algorithm performs a depth-first tree search. In its most efficient realization based on the Schnorr-Euchner enumeration of candidate paths [SE94], the best candidate is detected, its PED calculated, and the path is expanded to the next tree level. After reaching a tree leaf, the search moves up through the tree levels and pursues other possible symbol vectors at each level. During the tree search, paths with PEDs bigger than the search radius defined by d^2 are ignored. The search radius is initially undefined (i. e. $d^2 \to \infty$), whenever a tree leaf is reached, the search radius is updated with the Euclidean distance of the new path [SE94]. The sphere search concludes, when no more unprocessed paths within the search sphere remain.

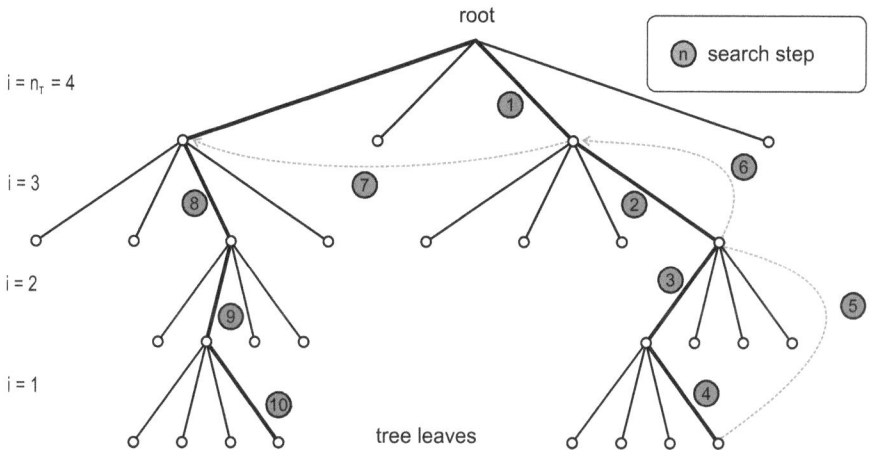

Figure 5.2: Sphere search in a 4×4 MIMO system for QPSK. The numbered search steps show the course of a possible SD tree search. Paths with thick lines have been pursued during the tree search; paths with thin lines have been excluded, because their PEDs are bigger than the sphere search radius.

The sphere search algorithm is guaranteed to find the ML solution with an average complexity significantly lower than that of a full search of the symbol space. However, the SD algorithm is a sequential algorithm and the number of required search iterations may vary. Mennenga et al. [MMF09] report the probability distribution of the sphere search complexity for the list sphere detector algorithm [HtB03], which computes further paths to determine soft-decision output. The probability distribution is of Gaussian shape with an average of approximately 45 searched paths and a significant variance (the exact value of the variance is not reported). For a real-time system with a fixed time budget, variable complexity means that the search needs to be terminated after a fixed number of searched paths, which potentially leads to a degraded performance.

Several modified sphere decoding algorithms that fix the number of searched paths and enable parallel processing have been proposed [BT06a, BT08b, WTW09, TSFB07, LBL+08, WTCM02, GM05, WEL09]. The principle approach is shown by two example algorithms in section 5.2: the K-best SD and the SSFE MIMO detector. Section 5.3 contains the description of the MIMO detection algorithm that has been implemented on the scalable SIMD processor architecture.

5.1.3 Soft-decision MIMO detection

The SD algorithm in the previous paragraph produces hard-decision output. A soft-output decoder computes the *a posteriori* probability (APP) of each signal bit based on the received signal vector $\mathbf{r}$. The probability of bit b_k can be expressed by the *log-likelihood ratio* (LLR) $L(b_k | \mathbf{r})$ (see equation (5.15)). The sign bit is the maximum a posteriori probability (MAP) estimate of b_k and the magnitude describes the reliability of the bit. LLRs are for example used in turbo decoders [BGT93] and LDPC decoders [Wib96].

$$L(b_k|\mathbf{r}) = \ln\frac{P(b_k=+1|\mathbf{r})}{P(b_k=-1|\mathbf{r})} \qquad (5.15)$$

For MIMO spatial multiplexing systems, the log-likelihood ratio can be estimated by equation (5.16) [HtB03, MZBF09].

$$L(b_k|\mathbf{r}) \approx \frac{1}{N_0} \cdot \min_{\mathbf{s}_{b_k=-1}} ||\mathbf{r} - \mathbf{H}\cdot\mathbf{s}||_2^2 - \frac{1}{N_0} \min_{\mathbf{s}_{b_k=+1}} ||\mathbf{r} - \mathbf{H}\cdot\mathbf{s}||_2^2 \qquad (5.16)$$

One of the squared Euclidean distance minima in equation (5.16) corresponds to the hard-decision minimum, the other value requires the processing of further tree paths. As the log-likelihood ratio has to be computed for each bit, computing the exact minima is too complex for a real-time implementation. Instead, usually only a limited list with good paths is computed. If no path for a bit value $b_k = +1$ or $b_k = -1$ is available[2], this bit value is apparently not probable and the corresponding minimum in equation (5.16) is replaced by an extreme value for the squared Euclidean distance. Soft-decision algorithms for the FSD MIMO detector are discussed below in section 5.3.3.

5.2 Breadth-first search MIMO decoders

All relevant SD algorithms in literature that fix the number of searched paths [BT06a, BT08b, WTW09, TSFB07, LBL+08, WTCM02, GM05, WEL09] replace the depth-first tree search of the original SD by a breadth-first tree search. This enables parallelizing the processing of paths, yet other criteria than the sphere radius are necessary to exclude paths and finally terminate the tree search. In the following, two examples are briefly discussed.

5.2.1 The K-best sphere decoder

The K-best sphere decoder limits the number of surviving paths at each tree level to K [WTCM02, GM05, GN06]. The decoder uses the M-algorithm for tree search after the QR-decomposition; therefore, it is also denoted as the QRD-M decoder. At each tree level, PEDs for all possible paths are computed and sorted. The K best paths are pursued in the next tree level, all other paths are discarded (see figure 5.3).

The main drawbacks of the K-best sphere decoder are the overhead for sorting PEDs, non-deterministic control flow, and the algorithm complexity $\mathcal{O}(K^{n_T})$, which depends on the number of surviving paths K at each level. Guo and Nilsson [GN06] compare the complexity and the BER of the K-best MIMO decoder using Schnorr-Euchner enumeration to the original Schnorr-Euchner SD [SE94] in a 4×4 MIMO system with 16-QAM. K-best decoder and SD achieve similar BERs for $K = 5$, yet the complexity of the K-best decoder is higher than the average complexity of the SD.

---

[2]The bit values -1 and 1 correspond to the binary values 1 and 0, respectively, the notation is based on BPSK modulation.

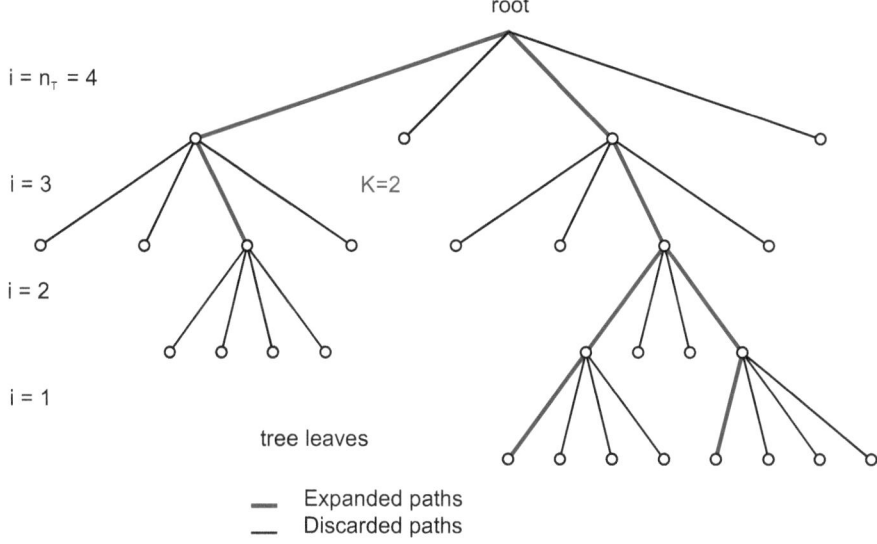

Figure 5.3: K-best tree search in a 4×4 MIMO system for QPSK: At each decoding level, the K paths with the best PEDs survive.

However, Joham et al. [JBL+08] demonstrate that the K necessary for close to ML performance can be reduced to values that are better manageable by using an unbiased MMSE metric.

5.2.2 Selective spanning with fast-enumeration

The selective spanning with fast-enumeration (SSFE) MIMO decoder [LBL+08, LFN+09, FLN+09] is another example of a breadth-first search sphere decoder with a fixed-number of paths. The SSFE algorithm is characterized by a vector $\mathbf{m} = [m_1, m_2, \ldots, m_{n_T}]$ which describes the number of surviving branches for each incoming path at each tree level; this part of the algorithm is called the selective spanning of nodes. The underlying concept originally has been introduced by Barbero and Thompson [BT06b]. The complexity of the tree search is $\mathcal{O}\left(\prod_{j=i}^{n_T} m_j\right)$. Exemplary search trees based on different spanning vectors are displayed in figure 5.4 for a 4×4 MIMO system.

Fast enumeration describes the algorithm for the enumeration of candidates at the tree levels. As the nodes in tree level i are processed independently (i.e. there is no need to compare PEDs from different nodes), it is not necessary to know the exact PEDs of all branches to determine the m_i

Chapter 5 Sphere decoding for MIMO detection

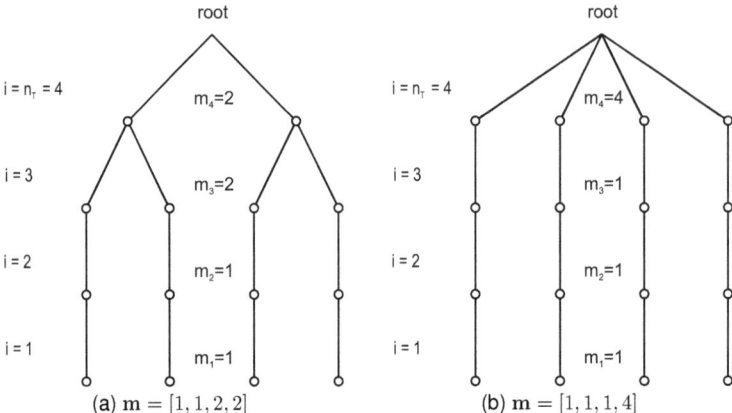

Figure 5.4: 4×4 MIMO SSFE search trees for two different spanning vectors

surviving paths. Instead of calculating PED increments $e_i(\mathbf{s})$ for all possible symbols, the received signal is mapped into the constellation space of symbol s_i:

$$
\begin{aligned}
e_i(\mathbf{s}) &= \left\| y_i - \sum_{j=i}^{n_T} R_{i,j} \cdot s_j \right\|_2^2 \\
\Leftrightarrow e_i(\mathbf{s}) &= \left\| y_i - \left(\sum_{j=i+1}^{n_T} R_{i,j} \cdot s_j \right) - R_{i,i} \cdot s_i \right\|_2^2 \\
\Leftrightarrow \frac{e_i(\mathbf{s})}{R_{i,i}} &= \left\| \underbrace{\left(y_i - \sum_{j=i+1}^{n_T} R_{i,j} \cdot s_j \right) / R_{i,i}}_{\xi_i} - s_i \right\|_2^2 \\
&= \| \xi_i - s_i \|_2^2
\end{aligned}
\quad (5.17)
$$

A geometrical interpretation for $m_i = 2$ is displayed in figure 5.5: Based on the computed value of ξ_i, the two closest constellation points are calculated and expanded to the next tree level. PEDs only have to be computed for these nodes.

According to Fasthuber et al. [FLN+09], near optimal detection performance can be achieved using the spanning vector $\mathbf{m} = [1, \ldots, 1, M]$ for modulation size M (e. g. as in figure 5.4b for QPSK modulation). For MIMO systems with up to four transmit and receive antennas, this spanning vector defines the same search tree as the tree of the FSD algorithm [BT06a, BT06c, BT08b]. Furthermore, the complete tree search is equivalent to the FSD algorithm, which is discussed in the following section.

Chapter 5 Sphere decoding for MIMO detection

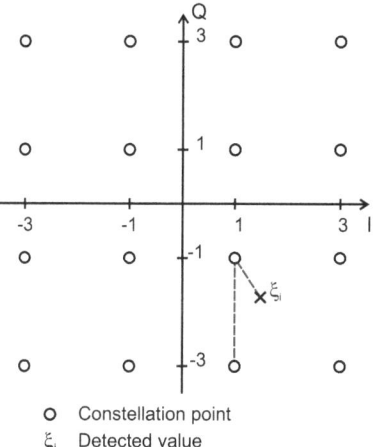

○ Constellation point
ξ_i Detected value

Figure 5.5: Fast enumeration of the two closest nodes based on distance to computed ξ_i for 16-QAM modulation

5.3 The fixed-complexity sphere decoder

The fixed-complexity sphere decoder (FSD) is a breadth-first tree search MIMO decoder, which enables parallel processing, fixes the number of visited tree nodes, and achieves quasi-ML performance. The algorithm has been proposed by Barbero and Thompson [BT06a, BT06c, BT08b] and mapped on a field-programmable gate array (FPGA) platform. It consists of two major parts: a novel channel matrix ordering based on noise amplification and a tree search through a fixed subset of the complete symbol space. Although, the channel matrix ordering has to be done before the QR-decomposition and the tree search, the requirements for the channel matrix ordering follow from the topology of the tree search. Hence, the tree search is explained first. Afterwards, the algorithm for the FSD ordering of the channel matrix is discussed.

5.3.1 FSD tree search

The FSD tree search is based on the ML detection, yet the number of searched candidates is limited to a subset $\mathcal{S}$:

$$\hat{\mathbf{s}}_{FSD} = \arg\min_{\mathbf{s} \in \mathcal{S}} ||\mathbf{r} - \mathbf{H} \cdot \mathbf{s}||_2^2 \tag{5.18}$$

The solution of equation (5.18) can be computed by a tree search starting with the n_T-th channel matrix row — evaluating only paths in $\mathcal{S}$. The subset of paths is created by fixing the number of branches in each tree level for a given modulation size M as follows:

- In the first p levels, all M branches are expanded. This part of the search is denoted as full expansion (FE).

- For the remaining levels of the tree, only one branch with the best PED metric is expanded, denoted as single expansion (SE).

The number of branches in each tree level can be summarized by the node distribution vector $\mathbf{n}_S = (n_1, n_2, \ldots, n_{n_T})^T$. The overall complexity of the tree search is $\mathcal{O}(M^p) \ll \mathcal{O}(M^{n_T})$. Barbero and Thompson [BT06b] show that quasi-ML performance can be obtained for 4×4 and 8×8 MIMO by setting the number of FE stages p to one ($\mathbf{n}_S = (1,1,1,M)^T$) and two ($\mathbf{n}_S = (1,1,1,1,1,1,M,M)^T$), respectively. Figure 5.6 shows the FE and SE stages of the tree search for $\mathbf{n}_S = (1,1,1,4)^T$ (corresponding to 4×4 MIMO with QPSK modulation). For 16-QAM, the performance degradation from the ML solution at $\mathrm{BER} = 10^{-3}$ is only 0.03 dB for a 4×4 and 0.25 dB for a 8×8 MIMO system and a Rayleigh fading channel. Jaldén et al. [JBOT09] also theoretically proved that the FSD tree search algorithm, combined with the channel ordering described below, achieves the same diversity as the ML detector for $p \geq \sqrt{n_T} - 1$. Furthermore, the FSD asymptomatically provides ML performance for high SNRs.

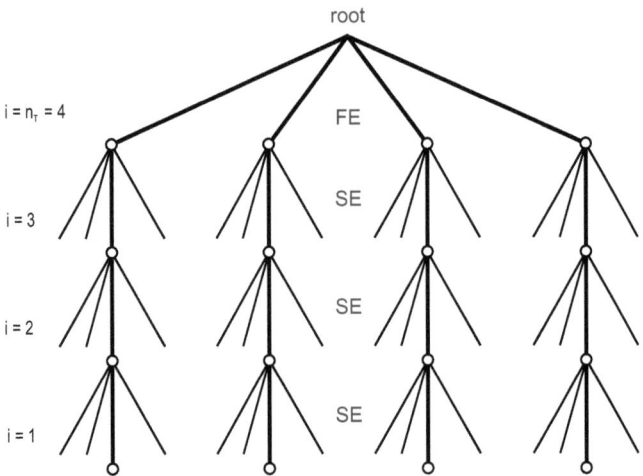

Figure 5.6: FSD tree search for QPSK and a 4×4 MIMO system: In full expansion (FE) stages all branches are evaluated, while only the best path is pursued in single expansion (SE) stages.

5.3.2 FSD ordering of the channel matrix

As shown by Jaldén et al. [JBOT09], the ordering of the channel matrix has a significant impact on the possible performance. The FSD channel matrix ordering algorithm reorders the matrix columns based on *noise amplification*; hence, the order in which the components of vector s are detected is changed. In FE stages, the noise amplification is irrelevant, as all possible paths are evaluated. Hence, the signals with the largest post-processing noise amplification should be processed in these

Chapter 5 Sphere decoding for MIMO detection

stages. In SE stages, high noise amplification may lead to detection errors, therefore signals with the smallest post-processing noise amplification should be processed in these stages.

The ordering of channel matrix columns, and hence the components of the transmitted signal vector, is done based on the properties of the product of the channel matrix with its conjugate transpose:

$$\mathbf{A} = \mathbf{H}^H \cdot \mathbf{H} \qquad (5.19)$$

The ordering algorithm iterates through the channel matrix columns starting with the last column (n_T). Each iteration has two steps:

1. The next signal component $\hat{s}_k$, which shall be detected, is determined by selecting the kth column of the channel matrix according to:

$$k = \begin{cases} \arg\max_j \left[\mathbf{A}^{-1}\right]_{j,j} & \text{if } n_i = M \\ \arg\min_j \left[\mathbf{A}^{-1}\right]_{j,j} & \text{if } n_i = 1 \end{cases} \qquad (5.20)$$

2. The matrix $\mathbf{A}$ is updated by removing its kth row and its kth column. This corresponds to removing the kth column from the channel matrix $\mathbf{H}$.

Although the FSD ordering algorithm requires a matrix inversion in each step, only the diagonal elements of the inverse matrix are required for solving equation (5.20). Hence, the complexity can be reduced by calculating just these elements.

5.3.3 Soft-decision MIMO detection based on the FSD

As explained in section 5.1.3, soft-decision sphere decoding is done by calculating a list containing candidate symbol vectors s and their Euclidean distance metrics. For each bit b_k, the difference between the Euclidean distances of the best candidate with $b_k = -1$ and the best candidate with $b_k = 1$ has to be computed (see equation (5.16)), if no candidate for a certain bit value is available, the Euclidean distance is assumed to be very big and the bit value is assumed to be unlikely. In order to prevent significant errors due to missing candidates, the number of candidates in the list should be increased compared to the hard-decision FSD, so that enough alternative paths with different bit values are available.

List FSD

Barbero and Thompson [BT08a] proposed a list extension of the FSD. The list FSD (LFSD) can also be described by the node distribution vector $\mathbf{n}_S = (n_1, n_2, \ldots, n_{n_T})^T$. $\mathbf{n}_S$ is initialized to the hard-decision FSD node distribution vector (e. g. $\mathbf{n}_S = (1, 1, 1, M)^T$ for 4×4 MIMO). Then, starting with tree level $i = n_T - p$ (the first single expansion stage) and proceeding down level by level, the number of evaluated branches at a tree level is doubled until a desired number of searched vectors N_S is attained. If the last level $i = 1$ is reached, the algorithm continues with level $i = n_T - p$. Figure 5.7 shows an example for QPSK modulation with node distribution vector $\mathbf{n}_S = (1, 1, 2, 4)^T$. Table

5.1 displays node distribution vectors, obtained from this algorithm, for different numbers of searched vectors in a 4×4 MIMO system with 16-QAM modulation. The list for the soft-decision detection is generated by selecting the $N_{\mathcal{L}}$ best paths ($N_{\mathcal{L}} \leq N_S$).

Table 5.1: Number of searched paths and node distribution vectors for the LFSD and a 4×4 MIMO system with 16-QAM modulation

Number of vectors	$N_S = 64$	$N_S = 128$	$N_S = 256$
Node distribution	$\mathbf{n}_S = (1,2,2,16)^T$	$\mathbf{n}_S = (2,2,2,16)^T$	$\mathbf{n}_S = (2,2,4,16)^T$

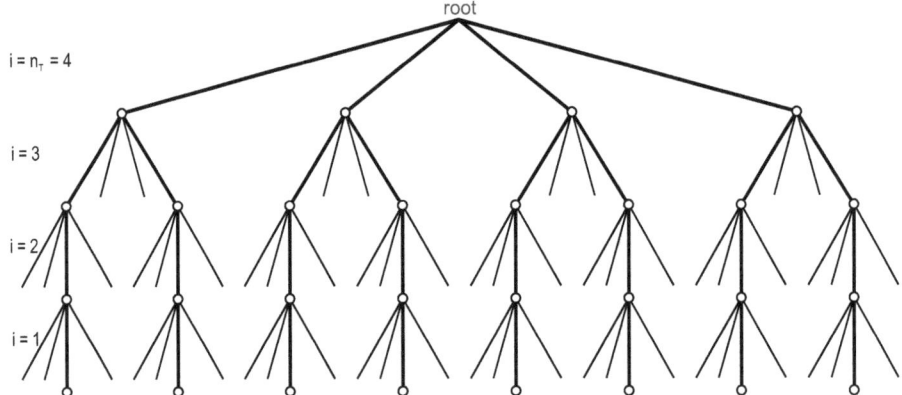

Figure 5.7: LFSD tree search for QPSK and a 4×4 MIMO system with $N_S = 8$

The main drawback of the proposed approach is the increased complexity of the tree search. In [BT08a], $N_S = 64$ is used for 4×4 MIMO and 16-QAM modulation, which means a quadrupling of the number of paths.

Smart candidate adding for QRD-M MIMO detection

Milliner, Zimmermann, Barry and Fettweis [MZBF09] proposed a different soft-decision MIMO symbol detection algorithm with a reduced complexity, which adds further candidates in a smart manner. The proposed algorithm is based on the M-algorithm, where the M best paths survive at each tree level (see section 5.2.1). The difference to the original M-algorithm is that further candidates are added at each tree level and the parameter M may change from tree level to tree level. At each tree level k, the proposed algorithm determines the node with the minimum PED, denoted as partial MAP node. For this node, additional candidates are generated by flipping bits: For a symbol bit width ω, ω new candidates are generated by flipping each of the bits of the symbol vector component s_k, the remaining symbol vector components $s_{k+1}, \ldots, s_{n_T}$ are left unchanged. After the expansion by bit-flipping, the number of paths may be pruned to the M-best paths, yet the newly added counter-hypotheses for the best path should be protected. This approach guarantees that at least one counter-hypothesis is available for each bit from symbol component s_k of the current best path. If the best path is the same at all tree levels, counter-hypotheses are available for each bit.

Smart candidate adding for the FSD algorithn

The bit flipping strategy can also be applied to the FSD algorithm. In each single expansion stage of the tree search, ω additional alternatives for the node with the smallest PED are generated by bit-flipping. A pruning of the candidate list, as done by the M-algorithm, is not performed. The resulting number of candidates depends on the number of antennas and the modulation size (see table 5.2). The topology of the bit-flipping soft-decision FSD tree depends on the PEDs, as additional candidates are only added for the best path at the current tree level. Figure 5.8 shows an example search tree for the case that the best path is the same at each tree level.

Table 5.2: Number of candidates for the soft-decision FSD with bit-flipping at each tree level for 4×4 MIMO

Modulation	Bit width	Candidates level 4	Candidates level 3	Candidates level 2	Candidates level 1
General case	ω	2^ω	$2^\omega + \omega$	$2^\omega + 2 \cdot \omega$	$2^\omega + 3 \cdot \omega$
QPSK	2	4	6	8	10
16-QAM	4	16	20	24	28

Compared to the hard-decision FSD, the bit-flipping soft-decision FSD variant leads to an increased processing complexity: The bit-flipping algorithm requires determining the best path at each SE stage of the algorithm. Hence, tree levels have to be processed sequentially and a minimum search is necessary. In the hard-decision case, there is no need for sequential processing of tree levels, i.e. one path can be processed from tree root to tree leaf before the next path is processed. Furthermore, soft-decision output requires keeping all paths in a list — including additional paths generated by bit-flipping. For hard-decision, it is sufficient to save the path minimum and the path that is currently processed. Additionally, the soft-decision bit-flipping FSD requires computing LLRs (see equation (5.16)) after the tree search has been finished.

5.4 SIMD implementation of the FSD for MIMO-OFDM

The FSD algorithm — as discussed in the previous section — has been implemented on the scalable SIMD processor architecture instead of the other presented algorithms for various reasons. As already discussed, the original sphere detector is a sequential algorithm with a variable decoding complexity. Therefore, the SD algorithm is ill suited for real-time processing on a SIMD processor, as e.g. shown by Mennenga et al. [MMF09].

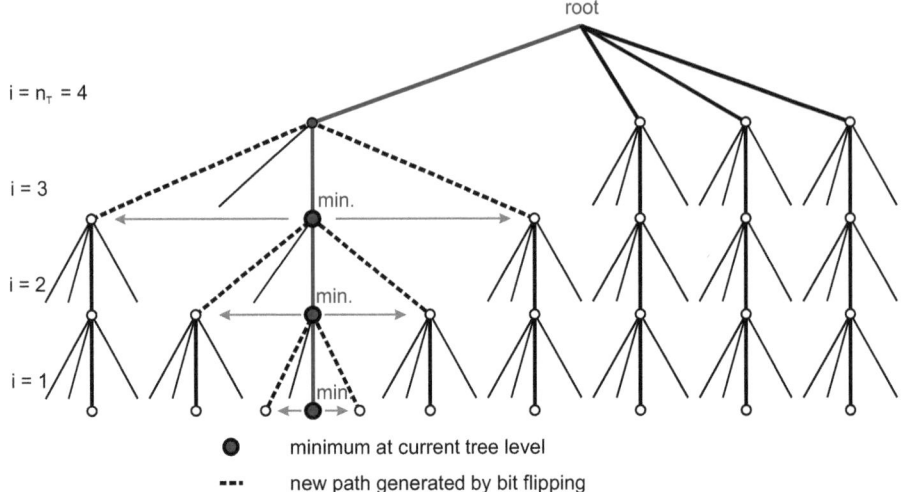

Figure 5.8: Example search tree for FSD with bit-flipping for generating additional paths (4×4 MIMO with QPSK modulation)

The other presented MIMO detection algorithms enable parallel processing by applying a breadth-first tree search strategy and fix the complexity of the tree search by limiting the number of candidates. The K-best sphere decoder, or QRD-M algorithm, calculates the K best paths at each tree level. Hence, the algorithm requires the sorting of PEDs, which may hinder efficient SIMD processing. Janhunen, Silvén and Juntii [JSJ09] report that the sorting prevents the efficient utilization of SIMD processing for the K-best sphere decoder on the Sandblaster SB3500 processor (see section 2.3.2). Both the SSFE and the FSD algorithm do not require a sorting of PEDs for hard-decision decoding[3]; hence, these algorithms can be more easily parallelized on a SIMD processor architecture. Both algorithms perform a similar tree search; in fact, the SSFE MIMO detector achieves the best performance if the FSD search tree is used [FLN+09]. Yet, the FSD algorithm also smartly reorders the channel matrix, which improves the diversity and the decoding performance of the system [JBOT09]. Hence, the FSD algorithm enables SIMD processing without sacrificing decoder performance.

The following paragraphs explain the implementation of the FSD algorithm for a 4×4 MIMO-OFDM system on the scalable SIMD processor architecture, as well as the encountered challenges. First, channel ordering and QR-decomposition are discussed. Then, the tree search for hard-decision output is explained. The last part of this section addresses the soft-decision tree search based on the bit-flipping algorithm for computing additional paths and the calculation of LLRs.

---

[3]The proposed soft-decision FSD algorithm with bit-flipping requires a minimum search in the single expansion stages of the algorithm.

5.4.1 Channel ordering

The FSD channel ordering requires calculating the matrix $\mathbf{A} = \mathbf{H}^H \cdot \mathbf{H}$. The channel ordering is then computed iteratively; each iteration requires a matrix inversion and a reduction of $\mathbf{A}$ by removing one row and one column (see equation (5.20)). Afterwards, the channel matrix needs to be reordered based on the new channel ordering. As the channel matrix is a 4×4 matrix and the iterations of the ordering algorithm lead to a further reduction of the matrix size, there are only limited opportunities for parallel processing for the channel ordering of one matrix. However, in an OFDM system, there is one channel matrix, one received vector, and one transmit vector for each OFDM sub-carrier. As the sub-carriers are orthogonal to each other, they may be processed in parallel during the channel ordering stage, as well as the QR-decomposition and the tree search. The principle approach is depicted by figure 5.9. Parallelism in this case is only limited by the number of data carriers in an OFDM symbol, which is sufficiently large for wide SIMD processing (e. g. 1200 data carriers for 20 MHz bandwidth in LTE).

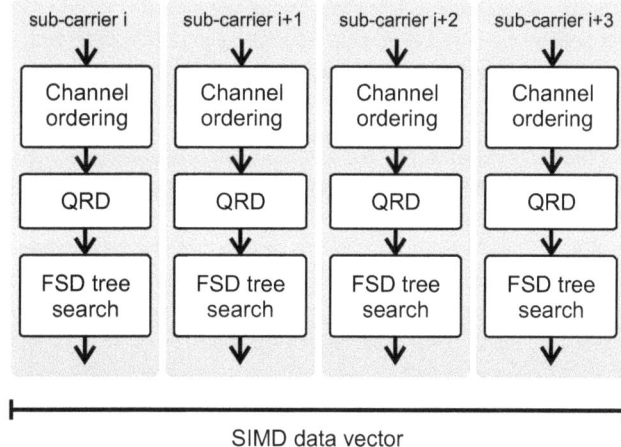

Figure 5.9: Parallel processing of the FSD algorithm by parallel processing of OFDM sub-carriers for a SIMD width of four elements

The main challenges of the channel ordering implementation are reducing the complexity of the matrix operations and avoiding roundoff errors due to the limited 16-bit fixed-point precision — especially during the matrix inversions.

Calculation of $\mathbf{A} = \mathbf{H}^H \cdot \mathbf{H}$

The first step of the channel ordering, the calculation of matrix $\mathbf{A}$ can be simplified due to the properties of the matrix. $\mathbf{A}$ is a Hermitian matrix, which means that the matrix is equal to its own conjugate transpose. The diagonal elements of $\mathbf{A}$ are real-valued and positive, as is the determinant of $\mathbf{A}$.

$$\mathbf{A} = \mathbf{H}^H \cdot \mathbf{H} = \begin{bmatrix} a_{1,1} & a_{2,1}^* & a_{3,1}^* & a_{4,1}^* \\ a_{2,1} & a_{2,2} & a_{3,2}^* & a_{42}^* \\ a_{3,1} & a_{3,2} & a_{3,3} & a_{4,3}^* \\ a_{4,1} & a_{4,2} & a_{4,3} & a_{4,4} \end{bmatrix} \quad a_{i,i} \in \mathbb{R}^+,\ a_{i,j} \in \mathbb{C} \forall i \neq j \quad (5.21)$$

Hence, only the diagonal elements and the elements below the diagonal need to be computed, the remaining six elements are given by symmetry. This reduces the computational complexity of the matrix product and the required memory for storing $\mathbf{A}$, as only ten matrix elements need to be saved. Furthermore, the property that diagonal elements are always real-valued allows to replace complex-valued multiplications by real-valued multiplications for operations involving one of these elements, which improves the performance, as a real-valued multiplication or MAC operation takes only one clock cycle, while a complex-valued operation requires two clock cycles.

Determining the channel ordering based on matrix-inversion

The next step of the channel ordering is calculating the index of the next channel matrix column from equation (5.20). The equation requires computing the position of the minimum or the maximum of the diagonal elements of $\mathbf{A}^{-1}$. Therefore, neither is there a need to compute the complete matrix inverse, nor a necessity to compute the *exact* values, as only the relative ordering of diagonal elements of the inverse matrix is required to determine the minimum or maximum position. Hence, equation (5.20) can be solved by computing the diagonal elements of the adjugate matrix of $\mathbf{A}$ instead of the inverse matrix:

$$\mathbf{A}^{-1} = \frac{1}{\det(\mathbf{A})} \cdot \operatorname{adj}(\mathbf{A})$$
$$\Rightarrow \arg\min_j \left[\mathbf{A}^{-1}\right]_{j,j} = \arg\min_j \left[\operatorname{adj}(\mathbf{A})\right]_{j,j} \quad (5.22)$$

Computing the diagonal elements of the adjugate matrix instead of the inverse of $\mathbf{A}$ reduces the computational complexity of the channel ordering, because a division by the matrix determinant can be avoided. Furthermore, the dynamic range of values is reduced, which reduces the impact of errors due to rounding and especially the saturation of values. The adjugate matrix of an $n \times n$ matrix is defined by the following equation, where $\mathbf{A}_{ij}$ denotes the sub-matrix that is generated by removing the ith row and jth column from $\mathbf{A}$:

$$\mathrm{adj}\,(\mathbf{A}) = \begin{bmatrix} \det(\mathbf{A}_{11}) & -\det(\mathbf{A}_{12}) & \cdots & (-1)^{n+1}\det(\mathbf{A}_{1n}) \\ -\det(\mathbf{A}_{21}) & \det(\mathbf{A}_{22}) & \cdots & (-1)^{n+2}\det(\mathbf{A}_{2n}) \\ \vdots & \vdots & & \vdots \\ (-1)^{n+1}\det(\mathbf{A}_{n1}) & (-1)^{n+2}\det(\mathbf{A}_{n2}) & \cdots & \det(\mathbf{A}_{nn}) \end{bmatrix} \quad (5.23)$$

Consequently, the channel ordering requires calculating and comparing four 3×3 determinants (4×4 input matrix), three 2×2 determinants (**A** reduced to 3×3), and one scalar comparison (**A** reduced to 2×2). The reduction of matrix **A** into smaller matrices by removing rows and columns for index k is visualized by figure 5.10. The required determinants can be directly computed, e.g. using the rule of Sarrus for 3×3 sub-matrices.

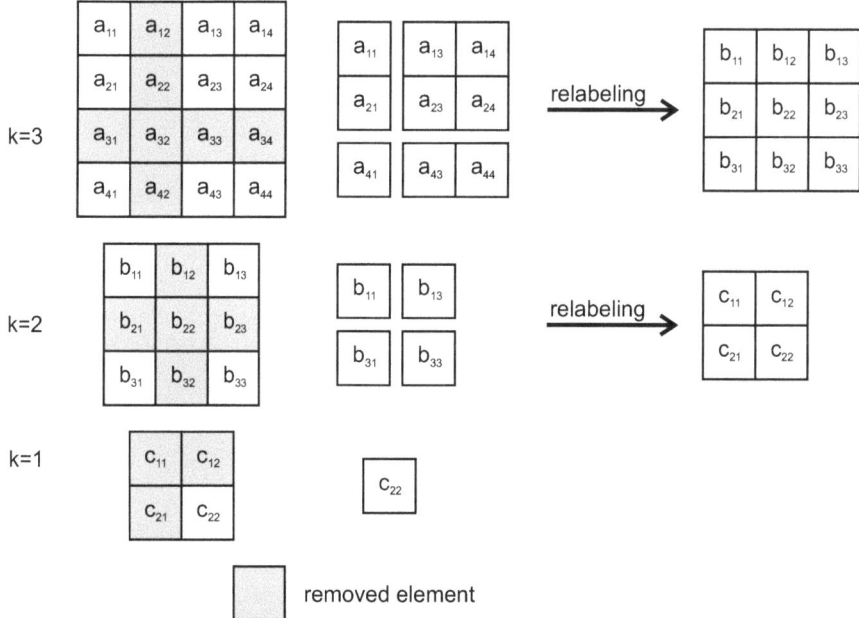

Figure 5.10: Example for the reduction of matrix **A** during the channel ordering. The relabeling of matrix elements is done to simplify the figure.

As the input data has a limited precision, due to the 16-bit word length, the proposed algorithm implementation has been tested for saturation or rounding errors. Rounding errors are insignificant, as the computed sub-matrix determinants are only used in comparison operations. Comparison errors will only occur if the values of two sub-matrix determinants are very close to each other. In this case, the relative ordering of the corresponding channel matrix columns is unimportant, as both columns suffer from a similar amount of post-processing noise-amplification.

Errors due to saturation are more significant, because a comparison of two saturated values is impossible. Furthermore, saturation of intermediate results (e. g. during the calculation of matrix $\mathbf{A}$) also leads to significant errors. Saturation can be avoided, by scaling the input values before or during the calculation of matrix $\mathbf{A}$ and by avoiding the division by a matrix determinant as in equation (5.22). Simulation results obtained from Matlab show that right shifting the input values by two bits is sufficient to prevent saturation under the assumption that the input channel matrix requires the full dynamic range provided by a 16-bit word length. If the channel estimation produces a channel matrix estimate with less than 15 bits precision, the scaling by right shifting may be avoided.

Reordering of channel matrix elements

The reordering of the channel matrix elements describes the removal of elements from $\mathbf{A}$, as in figure 5.10, and the ordering of the columns of the channel matrix $\mathbf{H}$ after the new column indices have been computed.

On a scalar processor architecture, both ordering operations can be efficiently implemented by memory access: The matrix elements are read from memory and stored in a different order, using the computed channel ordering for offset addressing. The overall complexity is one memory read and one memory write access per matrix element.

On a SIMD processor architecture, a different approach is necessary, as multiple OFDM sub-carriers — with potentially differing orderings — are processed in parallel in a vector. Hence, the ordering has to be done by conditionally swapping elements of two vectors, with masks defining the desired order of the channel matrix columns. The ordering of one matrix row with four elements requires six consecutive swapping operations, each swapping operation can be realized by a pair of parallel vector move operations (see figure 5.11).

```
vmov_vmac v0 v1 m1 || vmov_valu v1 v0 m1
```

Figure 5.11: Swapping of data vectors v0 and v1 based on vector mask m1 using two parallel masked move operations

Yet, although the channel reordering on vectors requires more operations than a channel reordering on scalars on the scalable SIMD processor architecture, the execution time is the same, as the overhead for swapping values can be efficiently hidden by LIW execution (see table 5.3). The runtime is determined by memory access operations.

Table 5.3: Complexity comparison of scalar and vector channel matrix reordering for a 4×4 matrix

Description	Load/Store operations	Swap operations	Runtime [Cycles]
Scalar: reordering by memory access	16+16	—	32
Vector: reordering by swapping values	16+16	24	32

5.4.2 QR-decomposition by Givens rotations

The next part of the FSD is the transformation of the (ordered) channel matrix into upper triangular form using a QR-decomposition by six successive complex-valued Givens rotations [GVL96]. A complex-valued givens matrix **G** is defined by:

$$\mathbf{G} \cdot \begin{bmatrix} a \\ b \end{bmatrix} = \begin{bmatrix} c^* & s^* \\ -s & c \end{bmatrix} \cdot \begin{bmatrix} a \\ b \end{bmatrix} = \begin{bmatrix} r \\ 0 \end{bmatrix} \quad (5.24)$$

$$c = \frac{a}{\sqrt{a^* \cdot a + b^* \cdot b}} \quad (5.25)$$

$$s = \frac{b}{\sqrt{a^* \cdot a + b^* \cdot b}} \quad (5.26)$$

$$r = \sqrt{a^* \cdot a + b^* \cdot b} \quad (5.27)$$

The required processing consists of two parts: the generation of the Givens rotation matrices and matrix products with the computed Givens matrices. The latter part can be easily realized by a series of complex-valued multiplications and MAC operations. Furthermore, each Givens rotation produces one real-valued output r; hence, further operations involving this matrix element can be performed using real-valued instead of complex-valued vector multiplications and MAC operations.

The generation of the Givens rotation matrices is more complicated, as it requires a reciprocal square root operation ($1/\sqrt{x}$), which is necessary to guarantee the numerical stability of the QR-decomposition.[4] In principle, the reciprocal square root can be computed using a lookup table (LUT), digit recurrence algorithms, iterative approximation by Newton-Raphson or Goldschmidt iterations, polynomial approximation, or a combination of multiple of these approaches. For a SIMD implementation, a lookup table is not a feasible approach. Jeannerod et al. [JKMR07] report that a polynomial approximation using a piecewise defined polynomial achieves the lowest cycle count for a 32-bit square-root operation on a LIW architecture compared to digit recurrence and iterative approximation algorithms. Hence, a reciprocal square root approximation based on a piecewise defined polynomial with two subintervals has been implemented for the Givens rotation algorithm. The implementation is a variation of the reciprocal square root algorithm used in the FLIP library for 32-bit floating-point calculations [Rai06].

---

[4]Square root (and division) free variants of the Givens rotation algorithm also exist [Gen73, GS91].

Reciprocal square root algorithm

As a first step, the input 16-bit fixed-point data x is transformed into a pseudo-floating point format $x = m_x \cdot 2^{e_x}$. The exponent is computed based on the number of leading zeros of the input value. The mantissa is then generated by shifting the input value, so that the value of the mantissa is between 0.5 and 1.0. The new mantissa m_y is calculated using two degree 3 polynomials:

$$m_y = \begin{cases} a_{3,0} \cdot m_x^3 + a_{2,0} \cdot m_x^2 + a_{1,0} \cdot m_x + a_{0,0} & \text{for } 0.5 \leq m_x < 0.75 \\ a_{3,1} \cdot m_x^3 + a_{2,1} \cdot m_x^2 + a_{1,1} \cdot m_x + a_{0,1} & \text{for } 0.75 \leq m_x < 1.0 \end{cases} \quad (5.28)$$

The polynomial coefficients are listed in table 5.4. In the implementation, the coefficients are all scaled by $1/4$ to map them on $Q.15$ fixed-point values. The correct polynomial can be selected by masked move operations. The polynomial evaluation can be realized by three MAC operations, one multiplication, and one shift operation to revert the pre-scaling of polynomial coefficients.

Table 5.4: Polynomial coefficients for reciprocal square root approximation

Input range for mantissa	a_3	a_2	a_1	a_0
$0.5 \leq m_x < 0.75$	1.9802...	-3.2286...	3.1369...	1.2011...
$0.75 \leq m_x < 1.0$	1.6636...	-1.9195...	1.3242...	0.3613...

Next, the new exponent e_y has to be calculated and the mantissa has to be updated based on the new exponent:

$$\begin{aligned} y &= \frac{1}{\sqrt{m_x}} \cdot \frac{1}{\sqrt{2^{e_x}}} \\ &= m_y \cdot 2^{-e_x/2} \\ &= m_y \cdot c_y \cdot 2^{e_y} \quad \text{with } e_y = \lfloor -e_x/2 \rfloor \end{aligned} \quad (5.29)$$

The correction term c_y is necessary to compensate for rounding errors during the calculation of the new exponent e_y. For even exponents no rounding error occurs, while compensation is necessary for odd exponents:

$$c_y = \begin{cases} 1 & \text{for even } e_x \\ \frac{1}{\sqrt{2}} & \text{for odd } e_x \end{cases} \quad (5.30)$$

SIMD parallelization of the QR-decomposition

Like the channel ordering, the QR-decomposition can only be efficiently vectorized on a wide SIMD processor architecture by processing multiple OFDM sub-carriers in parallel. Furthermore, the performance of the QR-decomposition can be improved by speeding up the reciprocal square root calculation, which is the most complex part of the QR-decomposition algorithm from an implementation point of view. A speedup can be achieved by calculating reciprocal square roots for two input vectors in parallel: The reciprocal square root operation is a real-valued operation, yet data vectors contain elements for the real and the imaginary part of a value. As only one of the two components is needed, another reciprocal square root operation can be performed in parallel by merging two data vectors. The principle approach is depicted by figure 5.12. The merging and the separation of data values from different vectors can be realized by simple masked permutation operations, e. g. 16-bit butterfly permutations or rotation operations.

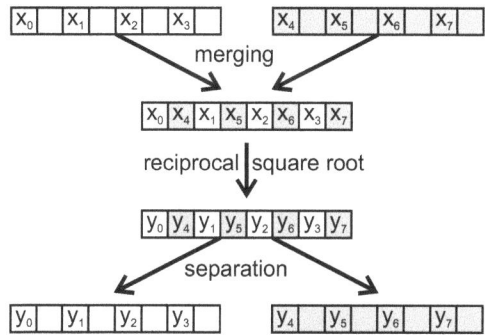

Figure 5.12: Merging of data vectors for the parallel computation of reciprocal square roots.

5.4.3 Hard-decision FSD tree search

The hard-decision FSD tree search can be parallelized by processing up to M paths in parallel (with M defining the modulation size, i.e. $M = 4$ for QPSK and $M = 16$ for 16-QAM) and by processing multiple OFDM sub-carriers in parallel. Due to the regular structure of the FSD tree, the latter approach produces no overhead at all and requires the least programming effort, as mostly the same assembly code can be used for all permutation network types and SIMD widths. Furthermore, as parallel processing of OFDM sub-carriers is also done during the channel ordering and the QR-decomposition, there is no need for data reordering. Hence, an implementation based on the parallel processing of OFDM sub-carriers has been realized. The processing of paths in the FSD tree is performed sequentially and the best path is always updated after reaching a tree leaf. The processing of the tree consists of the processing of the full expansion stage (symbol s_4) and the single expansion stages (symbols s_3, s_2, and s_1), as discussed below.

Full expansion stage

In the full expansion stage, all possible values of s_4 have to be sequentially enumerated and the PED $d_4 = ||y_4 - R_{4,4} \cdot s_4||_2^2$ has to be calculated. The enumeration of symbols has been realized by storing all symbol values in a vector (two vectors for 16-QAM and a 128-bit SIMD bit width) in a compressed form and rotating the symbol vector between tree search iterations.

Single expansion stages

In single expansion stages, only the best path survives. Therefore, there is no need to calculate all PEDs. Instead, the best candidate can be determined by mapping the received signal vector component r_i into the constellation space of symbol s_i, similar to the fast enumeration in the SSFE algorithm (see section 5.2.2). Based on equation (5.17), the best candidate for s_i can be computed as:

$$s_i = \arg\min_{s_k \in M} \left\| \left(y_i - \sum_{j=i+1}^{n_T} R_{i,j} \cdot s_j \right) / R_{i,i} - s_k \right\|_2^2 \tag{5.31}$$

$$\Leftrightarrow s_i = \arg\min_{s_k \in M} \left\| \underbrace{\left(y_i - \sum_{j=i+1}^{n_T} R_{i,j} \cdot s_j \right)}_{\chi_i} - R_{i,i} \cdot s_k \right\|_2^2 \tag{5.32}$$

Here, s_k enumerates all possible symbol values. The second equation avoids the costly division by $R_{i,i}$.

For rectangular modulation schemes, such as the QPSK and 16-QAM modulation schemes used in LTE [Tec09b], the distance minimization can be done in parallel for the in-phase and quadrature symbol components by one or multiple threshold comparisons. Figure 5.13 visualizes the approach for 16-QAM modulation. Assuming in-phase and quadrature symbol component values in $[-3, -1, 1, 3]$[5], the detection of the closest symbol can be performed by two threshold comparisons[6]:

$$\operatorname{Re}\{\chi_i\} \geq 0, \qquad\qquad \operatorname{Im}\{\chi_i\} \geq 0 \tag{5.33}$$
$$|\operatorname{Re}\{\chi_i\}| \geq 2 \cdot R_{i,i}, \qquad\qquad |\operatorname{Im}\{\chi_i\}| \geq 2 \cdot R_{i,i} \tag{5.34}$$

For QPSK, only the sign bit of χ_i has to be detected. After detecting the symbol component s_i in this manner, the algorithm proceeds to the next single expansion stage and the PED for the current path is updated.

5.4.4 Soft-decision FSD tree search extension by bit-flipping

The soft-decision FSD tree search extension by bit-flipping (see section 5.3.3) requires keeping a list of candidate paths, instead of only retaining the best path. Furthermore, all nodes and PEDs at a tree level i have to be computed before proceeding to the next tree level $i - 1$ and the current minimum path has to be computed to determine where additional paths have to be added by bit-flipping. The candidate list leads to an increased memory requirement for storing 28 (16-QAM) or 10 (QPSK) paths instead of one best path. The sequential processing of tree levels leads to a

---
[5]This is a simplification, which is done for better illustration of the approach.
[6]Note that after the QR-decomposition, the diagonal matrix elements $R_{i,i}$ with $i \in [1,2,3]$ are real-valued and positive.

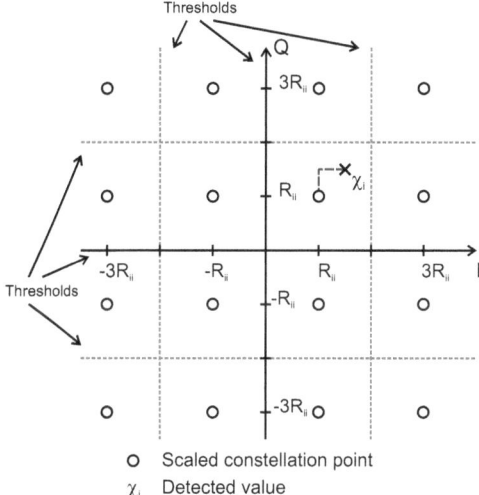

Figure 5.13: Symbol detection by threshold comparisons in the scaled constellation diagram for 16-QAM

loss in performance compared to the hard-decision FSD implementation, where complete tree paths are processed sequentially: During the hard-decision FSD tree search, paths are expanded by the described thresholding approach, the PEDs are only needed at the tree leafs for updating the best path. Hence, the PED calculation for a tree level can be done in parallel to the processing of the next tree level, reducing the overhead for computing the Euclidean distance norm. During the bit-flipping soft-decision FSD, PEDs are needed at each tree level to compute the candidate for bit-flipping.

Bit-flipping

The flipping of sign bits for QPSK and 16-QAM can be realized by one masked vector negation for the in-phase component and one masked vector negation for the quadrature component. For 16-QAM, a flipping of the amplitude bits is also required. Assuming that the valid symbol amplitudes are sym_1 and sym_2, the amplitude flipping can be realized by pre-computing $\tilde{s} = sym_1 \oplus sym_2$ and performing one masked exclusive or operation with $\tilde{s}$ for the in-phase component and one for the quadrature component. Hence, the bit-flipping requires only one clock cycle per flipped bit and vector.

5.4.5 LLR calculation for soft-decision MIMO decoding

After generating a list of candidates during the tree search with bit-flipping, LLR values have to be computed from the list of candidate symbol vectors and Euclidean distances. The output of the soft-decision tree search is the hard-decision minimum (defined by symbol vector s_{min} and squared

Euclidean distance metric d_{min}) and the unsorted list of candidates $\mathcal{L}$. The LLR for bit b_k (see equation (5.16)) can be estimated by computing:

$$L(b_k|\mathbf{r}) \approx \frac{1}{N_0} \cdot (-1)^{(\mathbf{s}_{\text{min},b_k}==1)} \cdot \left(d_{\text{min}} - \min_{\substack{\mathbf{s} \in \mathcal{L} \\ \mathbf{s}_{b_k} \neq \mathbf{s}_{\text{min},b_k}}} \|\mathbf{r} - \mathbf{H} \cdot \mathbf{s}\|_2^2\right) \quad (5.35)$$

Each LLR calculation requires one subtraction, one masked negation (based on the bit value of $\mathbf{s}_{\text{min},b_k}$), a multiplication by $1/N_0$, and a minimum search on list elements, whose bit b_k differs from the hard-decision minimum. The minimum search is the most complex task, as the list is not sorted and the bit values are not ordered. Hence, all elements in the list have to be compared with $\mathbf{s}_{\text{min}}$ and the minimum has to be updated based on the comparison result[7]. The comparison with $\mathbf{s}_{\text{min}}$ can be efficiently parallelized by storing all symbol bits in one vector element[8] and performing a bitwise exclusive or operation ($\mathbf{s}_{\text{min}} \oplus \mathbf{s}$). The resulting bit values are then converted into vector masks and utilized to update the minimum with a masked minimum operation. The proposed minimum update algorithm requires nine clock cycles per list candidate for 16-QAM modulation (one exclusive or and eight minimum operations on the VALU) and five clock cycles per list candidate for QPSK modulation (one exclusive or and four minimum operations on the VALU).

5.5 Performance analysis

In the following, the performance of the FSD on the scalable SIMD processor architecture is analyzed. First, the overall efficiency of the different FSD algorithm parts is assessed. Next, the throughput of the hard-decision and soft-decision FSD implementations is evaluated based on the requirements for future 4G wireless systems. The analysis concludes with a performance comparison to other SDR, ASIC, and FPGA implementations of similar sphere-based MIMO detectors. At the end of this section, approaches to improve the algorithm performance with custom instructions are discussed.

5.5.1 Overview of FSD results

Table 5.5 gives an overview of the implementation results on the scalable SIMD processor architecture and — for comparison purposes — on the EVP (see section 2.3.1). Columns two and three display the runtime of the different algorithm parts in clock cycles per vector on the scalable SIMD processor architecture and the EVP respectively. The QR-decomposition loop processes two vectors in parallel and requires 421 clock cycles on the SIMD architecture, which explains the normalized runtime of 210.5 clock cycles. The runtime per processed channel matrix depends on the number of 32-bit elements (16-bit imaginary part, 16-bit real part) in a vector. The runtime does not depend on the SIMD width. Hence, linear speedups can be obtained by increasing the SIMD width.

---

[7] The initial minimum is set to the maximum 16-bit integer value (32767).
[8] In-phase and quadrature components are still stored separately.

The runtimes on the scalable SIMD processor architecture are lower than the runtimes on the EVP, although the EVP potentially offers more LIW parallelism and a slightly richer instruction set[9]. The difference results from the used programming approach: The EVP is programmed in EVP-C, a C-based language with extensions for vector operations; the scheduling of operations and the register assignment is done by a compiler. On the other hand, the scalable SIMD processor architecture is programmed in assembly code, which leads to a more efficient scheduling of instructions. The soft-decision FSD algorithm with bit-flipping has not been implemented on the EVP, as a license for the software development tools has not been available.

Table 5.5 also shows the average number of parallel operations in a LIW instruction ($N_{\text{par. }\varnothing}$) for the inner loops of the algorithm parts, as well as the resource utilization of the vector MAC unit (VMAC, R_{VMAC}) and the vector ALU (VALU, R_{VALU}). The soft-decision FSD tree search consists of multiple loops; hence, average values for $N_{\text{par. }\varnothing}$, R_{VMAC}, and R_{VALU} are shown. LIW parallelism does not depend on the SIMD width, yet there are minimal differences (± 1 vector permutation operation per loop iteration) for permutation networks with different widths for the soft- and hard-decision tree search for 16-QAM. For these algorithm parts, the average value $N_{\text{par. }\varnothing}$ for all permutation networks is listed.

Table 5.5: Overview of FSD implementation results. Columns two and three show the runtime in clock cycles per vector on the scalable SIMD architecture and the EVP. The remaining columns contain the average number of parallel operations per instruction ($N_{\text{par. }\varnothing}$) and the VMAC and VALU resource utilization.

Algorithm part	SIMD pro. [Cycles/vector]	EVP	$N_{\text{par. }\varnothing}$	R_{VMAC} [%]	R_{VALU} [%]
Channel ordering	193	203	1.89	88.12	47.67
QR-decomposition	210.5	250	1.78	91.21	32.54
Hard-decision FSD search (QPSK)	127	148	2.125	89.29	60.71
Hard-decision FSD search (16-QAM)	503	640	2.61	87.30	69.84
Soft-decision FSD search (QPSK)	231	—	2.80	76.67	83.33
LLR calculation (QPSK)	65	—	3.40	0.00	100
Soft-decision FSD search (16-QAM)	855	—	2.76	78.44	82.54
LLR calculation (16-QAM)	281	—	3.22	0.00	100

Both the channel ordering and the QR-decomposition utilize the VMAC almost all the time, as these algorithms are based on matrix multiplications on complex-valued matrices. The high resource utilization values indicate the efficiency of the implementation; higher values for R_{VMAC} have not been obtained, because some algorithm parts do not require the VMAC. During the channel ordering, the VALU is mostly used for scaling operations (to avoid errors due to the fixed word length) and the computation of the new channel ordering. The QR-decomposition algorithm requires operations on the VALU for the calculation of Givens rotation matrices. The average number of parallel operations

---
[9]For example, the scalable SIMD processor architecture supports vector move operations on three different units, while the EVP supports move operations on any vector unit.

per instruction is relatively low, because most of the time, only the VMAC is used for complex-valued MAC operations and multiplications. The instruction count is low, because complex-valued operations require two clock cycles.

In the FSD tree search implementation, the VMAC is mostly utilized for PED calculations, while the VALU is required for the symbol candidate selection, the minimum update, and the bit-flipping in the soft-decision FSD. During the hard-decision tree search, PED calculations on the VMAC and symbol selection operations on the VALU can be efficiently parallelized, leading to a low runtime and a high VMAC utilization. The soft-decision FSD with bit-flipping requires sorting at each tree level, which increases the number of operations on the VALU and prevents an efficient parallelization of PED calculations and symbol selection operations. Hence, the VALU resource utilization increases, while the VMAC resource utilization decreases.

The soft-decision LLR calculation implementations achieve a very high degree of LIW parallelism and a 100 percent utilization of the VALU for masked minimum operations and exclusive or operations. The VMAC unit is not required for the calculation of the best paths for a desired bit-value.

5.5.2 Analysis of the achievable throughput

The FSD throughput can be calculated from the number of detected transmit vectors per second, with each transmit vector containing four elements with four (16-QAM) and two (QPSK) data bits. Table 5.6 lists the achieved throughputs on a 128-bit SIMD processor architecture at a clock frequency of 300 MHz for hard-decision and soft-decision decoding. Two different throughputs are listed: the worst-case throughput and the best-case throughput.

Table 5.6: 4×4 FSD throughput on 128-bit SIMD processor architectures

Algorithm	Worst-case [Mbps]	Best-case [Mbps]
Hard-decision FSD, QPSK modulation	18.10	54.86
Hard-decision FSD, 16-QAM modulation	21.18	34.85
Soft-decision FSD, QPSK modulation	13.72	28.91
Soft-decision FSD, 16-QAM modulation	12.47	16.21

The worst-case calculation assumes a fast fading channel, i.e. the channel matrix coefficients change during the transmission of one frame. Hence, the channel ordering and the QR-decomposition have to be recomputed for each OFDM symbol. Therefore, the runtime is calculated by adding up the channel ordering, QR-decomposition, and sphere search runtime (and — for soft-decision MIMO detection — the LLR calculation runtime).

The best-case throughput calculation assumes a slow fading channel, i.e. the channel matrix is assumed constant for the duration of one frame. In this case, the runtime for channel ordering and QR-decomposition can be neglected, only the transformation of the received vector based on the Q matrix has to be computed:

$$\mathbf{y} = \mathbf{Q}^H \cdot \mathbf{r} \qquad (5.36)$$

Chapter 5 Sphere decoding for MIMO detection

In the following, the best-case results are used for the throughput, as this is the usual procedure in literature and MIMO transmission on a slow fading channel is more realistic than MIMO transmission on a fast fading channel.

Figure 5.14 displays the (best-case) throughput for hard-decision and soft-decision MIMO detection for different SIMD widths in a logarithmic scale. The throughput scales linearly with the SIMD width. For hard-decision MIMO detection (figure 5.14a), a maximum throughput of 438.86 Mbps for QPSK and 278.77 Mbps for 16-QAM is achieved on 1024-bit SIMD processors. The soft-decision throughput is approximately half as much, with a maximum of 223.26 Mbps for QPSK and 129.73 Mbps for 16-QAM on a 1024-bit SIMD processor. The soft-decision throughput is significantly lower than the hard-decision throughput, because additional paths have to be calculated by bit-flipping and because of the additional overhead for the LLR calculation.

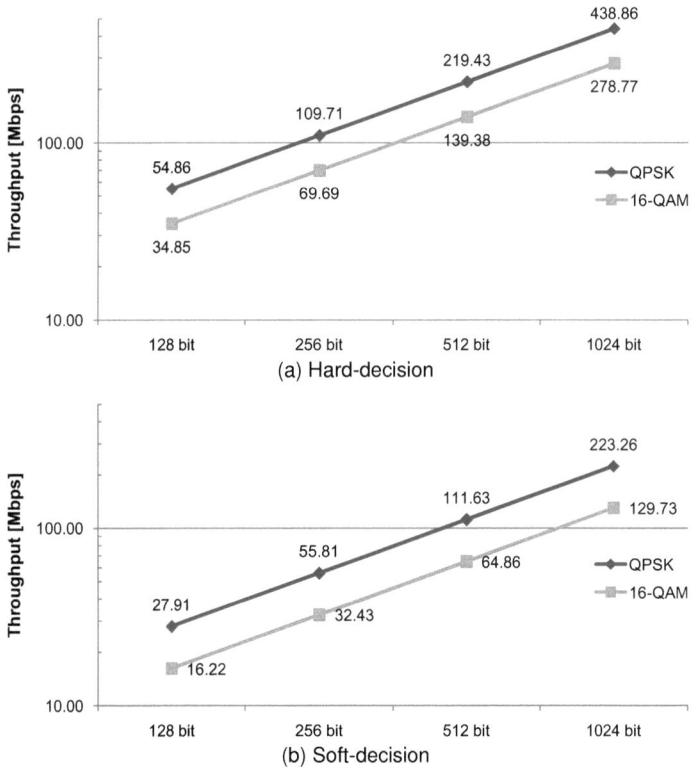

Figure 5.14: Best-case 4×4 MIMO FSD throughput for different SIMD widths for hard-decision (a) and soft-decision (b) decoding

Throughput requirements for 4×4 MIMO based on the LTE frame structure

The future 4G standard LTE-Advanced is based on LTE. Therefore, the following throughput analysis assumes a frame structure based on LTE with 4×4 MIMO support for higher data rates. An LTE frame has a duration of 10 ms [Tec09b], each frame consists of 20 slots. In short CP mode, seven OFDM symbols are transmitted per slot. Assuming a bandwidth of 20 MHz, one OFDM symbol comprises 2048 sub-carriers, with 1200 of these carriers containing data. The required maximum throughput is given by:

$$\text{Throughput}_{\text{max}} = \frac{N_{\text{FFT/slot}} \cdot N_{\text{data carrier}} \cdot 4 \cdot \log_2(M)}{t_{\text{slot}}} \quad (5.37)$$

For the assumed frame structure, a maximum throughput of 134.4 Mbps is required for QPSK modulation ($M = 4$) and a throughput of 268.8 Mbps is required for 16-QAM modulation ($M = 16$).

The implemented hard-decision FSD algorithm can meet these requirements. The required throughput for QPSK modulation can be reached on a single 512-bit processor running at a significantly reduced clock frequency of less than 185 MHz, the required throughput for 16-QAM modulation can be reached on a 1024-bit SIMD processor running at approximately 290 MHz. The required performance can also be reached by using multiple smaller SIMD processors.

The bit-flipping soft-decision FSD algorithm achieves a significantly lower throughput. A 1024-bit SIMD processor is required for QPSK modulation (at a clock frequency of approximately 180 MHz). Soft-decision FSD decoding for 16-QAM is not possible on a single SIMD processor core for a 20 MHz channel with 1200 data carriers. However, these are only the throughput requirements for the maximum channel bandwidth and the case that all data carriers are allocated to one user. Other channel bandwidths, with different numbers of data carriers are also supported by LTE [Tec10]. Table 5.7 lists the parameters for different channel bandwidths and the required SIMD width to decode all data carriers for 16-QAM modulation. Table 5.8 lists the maximum number of resource blocks that can be decoded in real-time; one resource block consists of 12 sub-carriers.

Table 5.7: Throughput requirements for different channel bandwidths for 4×4 MIMO with 16-QAM modulation

Bandwidth	FFT size	Data carriers	Required throughput	Min. SIMD width
1.4 MHz	128	72	16.13 Mbps	128 bit
3 MHz	256	180	40.32 Mbps	256 bit (hard.) / 512 bit (soft.)
5 MHz	512	300	67.2 Mbps	256 bit (hard.) / 1024 bit (soft.)
10 MHz	1024	600	134.4 Mbps	512 bit (hard.)
15 MHz	1536	900	201.6 Mbps	1024 bit (hard.)
20 MHz	2048	1200	268.8 Mbps	1024 bit (hard.)

Table 5.8: Maximum number of resource blocks that can be decoded in real-time for 4×4 MIMO with 16-QAM modulation

SIMD width	Resource blocks	Data carriers	Resource blocks	Data carriers
	Hard-decision FSD		Soft-decision FSD	
128 bit	12	144	6	72
256 bit	25	300	12	144
512 bit	51	612	24	288
1024 bit	103	1236	48	576

5.5.3 Comparison to SDR and hardware-based sphere decoders

Table 5.9 lists sphere decoders implemented as ASICs or on FPGA. The implemented sphere decoders all apply a breadth-first search strategy with a fixed search tree. Guo and Nilsson [GZMC06] designed an ASIC based on a modified K-best sphere decoder ($K = 5$) using the Schnorr-Euchner enumeration for soft-decision sphere decoding. Barbero and Thompson [BT06c] implemented the FSD algorithm on a Xilinx FPGA. Fasthuber et al. proposed a scalable ASIC for soft-decision MIMO detection based on the SSFE algorithm, the search tree is the same as the FSD tree.

Table 5.9: Performance of sphere decoding algorithms with fixed-complexity in ASICs and on FPGA

	[GN06]	[BT06c]	[FLN+09]
MIMO system		4×4	
Modulation		16-QAM	
Algorithm	Modified K-best (soft-decision, $K = 5$)	FSD	SSFE (soft-decision, m = $(1, 1, 1, 16)$)
Platform	ASIC ($1.07\,\text{mm}^2$ @ $0.13\,\mu$m)	Xilinx XC2VP70 FPGA	ASIC ($0.29\,\text{mm}^2$ @ 65 nm)
Clk. frequency	200 MHz	100 MHz	400 MHz
Throughput	106.6 Mbps	400 Mbps	400 Mbps
Power	—	—	28 mW

The proposed implementation on the scalable SIMD processor architecture achieves a better soft-decision throughput (129.73 Mbps on a 1024-bit SIMD processor) than the K-best sphere decoder and about two-thirds of the throughput of the FPGA FSD implementation (278.77 Mbps for 1024-bit SIMD width) for hard-decision decoding.

The MIMO detection ASIC based on the SSFE algorithm by Fasthuber et al.[FLN+09] offers the highest throughput with very low area and power consumption. This chip clearly defines the state-of-the-art for dedicated MIMO detection accelerators. Compared to the 1024-bit SIMD implementation, the soft-decision throughput is approximately three times higher on the SSFE ASIC. The SIMD processor requires approximately 10 times as much power and area as the ASIC.

Table 5.10 lists performance figures for MIMO detection algorithms on programmable processor architectures for SDR. Li, Bougard et al. [LBL+08] implemented the SSFE algorithm for spanning vector $\mathbf{m} = (1, 1, 2, 4)^T$ on a high-performance fixed-point DSP core. The TMS320C6416 processor is a LIW processor with six ALUs and two multipliers. Each unit can perform two 16-bit operations in parallel. The implementation realizes hard-decision MIMO detection for 4×4 MIMO and 64-QAM modulation.

Table 5.10: Performance of sphere decoding algorithms with fixed-complexity on SDR processors

	[LBL+08]	[LFN+09]	[JSJ09]
MIMO system	4×4	2×2	
Modulation	64-QAM	16-QAM	64-QAM
Algorithm	SSFE ($\mathbf{m} = (1,1,2,4)$)	SSFE (soft-decision, unknown m)	K-best ($K = 8$)
Processor	TMS320C6416 VLIW DSP	ASIP based on ADRES (2×4 256-bit SIMD)	SB3500 (3 SBX cores)
Clk. frequency	720 MHz	400 MHz	600 MHz
Throughput	37.4 Mbps	368 Mbps	3.42 Mbps
Power	$\approx$1.7 W [Hie03]	—	$\approx$360 mW

Li, Fasthuber et al. [LFN+09] implemented the SSFE algorithm for soft-decision MIMO detection on an ASIP based on the ADRES reconfigurable array (e. g. [MVV+03, SVPG+10]) with a 2×4 array configuration and 256-bit SIMD processing units as array elements and application-specific instructions. The used spanning vector is not reported, the results for 16-QAM modulation assume a 2×2 MIMO system.

Janhunen, Silvén and Juntii [JSJ09] implemented a K-best sphere decoder for $K = 8$ on the SB3500 processor (see section 2.3.2) for 2×2 MIMO and 64-QAM modulation.

The proposed implementation of the FSD algorithm outperforms the SSFE implementation on the TMS320C6416 processor and the K-best sphere decoder on the SB3500 processor, while the SSFE implementation on the ASIP based on ADRES achieves apparently a higher throughput. Yet, the throughput results are only for 2×2 MIMO, which is significantly less complex than 4×4 MIMO. Fasthuber et al. [FLN+09] report that there is a quadratic growth in complexity for the LLR calculation from 2×2 to 4×4 for the SSFE algorithm; the complexity of the tree level processing grows linearly. Hence, the performance for 4×4 MIMO will be significantly lower. Furthermore, the high throughput can only be reached with eight 256-bit SIMD processors and application-specific instruction set extensions.

If eight 256-bit SIMD cores are available for soft-decision MIMO detection based on the FSD, a throughput of 259.46 Mbps can be reached for a MIMO system with *four* receive and transmit antennas.

5.5.4 Improving the FSD performance

The performance results for the FSD algorithm on the proposed scalable SIMD processor architecture have been obtained without any application-specific modifications or extensions to the processor architecture. Yet, during the implementation of the FSD algorithm, performance bottlenecks that can be fixed with small extensions to the instruction set have been identified. Below, these bottlenecks and potential solutions are briefly discussed.

The calculation of the squared Euclidean norm ($\|\cdot\|_2^2$) is a major performance bottleneck, as the calculation requires multiple operations and lies in the critical path of the FSD algorithm. The hard-decision FSD algorithm requires at least 16 squared Euclidean norm computations and the soft-decision algorithm requires at least 72 squared Euclidean norm computations for 16-QAM modulation. Each norm computation requires three consecutive operations on the scalable SIMD processor architecture (see figure 5.15): First the real and imaginary parts are squared using a multiplication operation. Next real and imaginary parts are swapped using a permutation operation. In the last step, both values are accumulated.

```
1: vmul_f16_rdnsat v0 v0 v0
2: vswap v1 v0
3: vaddsat v0 v0 v1
```

Figure 5.15: Assembly code fragment for the calculation of the squared Euclidean distance

A specialized squared Euclidean norm operation could possibly perform the same operation on the VMAC unit in one clock cycle by first squaring real and imaginary parts using the multipliers and then accumulating the results for real part and imaginary part.

The computation of the best symbol candidate by thresholding (see figure 5.13) during the single expansion stages of the FSD is a further performance bottleneck, especially for 16-QAM modulation. Assembly code for 16-QAM thresholding is displayed in figure 5.16: First, the absolute value of the input vector v0 is calculated and the symbol value sym_2 is broadcasted to v4. Next, the absolute value is compared to the threshold for the amplitude of the symbol vector and sym_1 is broadcasted to v5. Then, the amplitude is updated, while the sign is determined by comparing v0 to zero. In the fourth clock cycle, the sign of the symbol vector is updated. The VALU is occupied all the time.

```
1: vabs v2 v0    || vbcst16 v4 r2
2: vcmpgte m1 v2 v1 || vbcst16 v5 r1
3: vcmpgte m2 v2 v3 || vmov_vmac v4 v5 m1
4: vneg v4 v4 m2
```

Figure 5.16: Assembly code fragment for the thresholding operation during SE stages for 16-QAM modulation

The thresholding is a performance bottleneck if no other useful operations can be done in parallel, e.g. PED computations on the VMAC. During the soft-decision FSD, PED computations cannot always be done in parallel to the thresholding; hence, the performance can be improved by speeding up the thresholding operation. The thresholding can be realized using small programmable lookup tables (LUTs) that are distributed to the 16-bit SIMD lanes. Each LUT contains the possible symbol values; the address is generated from the MSBs of the data values.

The thresholding requires two operations on the VALU for QPSK modulation; hence, there is less room for improvement. Yet, QPSK thresholding could also be implemented using small LUTs.

A further performance bottleneck is the minimum-search during the LLR calculation for the soft-decision FSD, which requires one vector minimum operation per pair of bits (in the in-phase and quadrature signal components) and soft-decision list element. The performance could be improved by computing the required minimum for multiple bits in parallel. As the channel decoding algorithm (e.g. a turbo decoder or an LDPC decoder) does not require the LLR values in 16-bit precision, the LLR calculation could be performed on 8-bit data types, with two 8-bit elements stored in one 16-bit vector element, potentially leading to a runtime reduction by 50 percent.

5.6 Conclusion

Sphere decoding can be efficiently realized on arbitrary wide SIMD processor architectures if two prerequisites are satisfied. Firstly, a breadth-first search strategy has to be applied to the tree search instead of the original sequential depth-first sphere search algorithm. The FSD algorithm fulfills this requirement and still achieves close to ML bit error rates. Secondly, parallel processing of multiple sphere searches has to be enabled, as parallelism in one sphere search is limited by the size of the modulation symbol alphabet (e.g. four different symbols for QPSK modulation). Future MIMO systems will probably use OFDM block modulation. Hence, this requirement is fulfilled, as orthogonal OFDM sub-carriers can be processed in parallel.

The hard-decision FSD implementation can meet the throughput requirements of 4×4 MIMO systems based on the LTE frame structure (up to 278.77 Mbps for 16-QAM modulation on a 1024-bit SIMD processor). Due to the significantly increased complexity, the implemented soft-decision FSD algorithm based on bit-flipping achieves approximately half the throughput of the hard-decision FSD algorithm.

The FSD implementation on the scalable SIMD processor architecture achieves approximately 32 percent of the soft-decision throughput of the best known hardware implementation [FLN+09], while also consuming more power and requiring more area. Compared to other SDR implementations, the achieved performance for both hard-decision and soft-decision MIMO detection is very good.

Chapter 6

Decoding of quasi-cyclic low density parity check codes

The decoding of quasi-cyclic low-density parity check (LDPC) codes on the proposed scalable SIMD processor architecture is evaluated in this chapter. Section 6.1 describes the basics of LDPC coding, such as the representation by Tanner graphs and the properties of quasi-cyclic LDPC codes. The following section (section 6.2) explains the decoding of LDPC codes by message-passing algorithms. SIMD implementations of WiMAX LDPC codes are discussed in section 6.3 and their performance is evaluated in section 6.4. Finally, conclusions are drawn in section 6.5.

6.1 Fundamentals

LDPC codes are parity check codes based on very sparse matrices. The BER performance of LDPC codes can be close to the Shannon limit [MN97, RSU01, CFRU01]. LDPC codes have been invented by Gallager and first published in his dissertation in 1960 [Gal63]. Yet, interest in LDPC codes only developed after the advent of turbo codes, which were invented by Berrou and Glavieux in 1993 [BGT93]. LDPC codes were finally rediscovered independently by MacKay and Neal [MN95] and by Wiberg [Wib96].

6.1.1 Definition of LDPC codes

LDPC codes are block codes that are defined by a parity check matrix $\mathbf{H}$ in $\mathrm{GF}\,(2)$.[1] $\mathbf{H}$ is a $(N-K) \times N$ matrix, N is the code word length and K is the number of information bits. The parity check matrix defines $M = N - K$ parity check bits. The code rate is $R = K/N$. A codeword $\mathbf{c} = [c_1, c_2, \ldots, c_N]$ is valid if a multiplication with the check matrix produces the zero vector:

$$\mathbf{H} \cdot \mathbf{c}^T = \mathbf{0} \qquad (6.1)$$

Each row of the parity check matrix defines a parity check condition.

---

[1]LDPC codes can also be defined in $\mathrm{GF}\,(q)$ [DM98, BD03], yet so far these codes have not been considered in wireless standards.

The encoding of a bit sequence $\mathbf{b} = [b_1, b_2, \ldots, b_K]$ is done by a product with a $K \times N$ generator matrix G:

$$\mathbf{c} = \mathbf{b} \cdot \mathbf{G} \tag{6.2}$$

The generator matrix can be calculated from the parity check matrix in systematic form, which can be obtained by elementary matrix transformations:

$$\tilde{\mathbf{H}} = \begin{pmatrix} -\mathbf{A}^T & \mathbf{I}_{N-K} \end{pmatrix} \qquad \mathbf{A} : K \times (N-K) \tag{6.3}$$

$$\mathbf{G} = \begin{pmatrix} \mathbf{I}_K & \mathbf{A} \end{pmatrix} \tag{6.4}$$

In general, the encoding of LDPC codes has a complexity of $\mathcal{O}(N^2)$, yet Richardson and Urbanke [RU01] developed a method based on the triangulation of the check matrix that enables encoding with almost linear complexity.

LDPC codes can be described by the Hamming weights of rows w_r and columns w_c. In the graphical representation by Tanner graphs, these values describe the check node and bit node degrees (see section 6.1.2). A sparse parity check matrix satisfies the conditions $w_r \ll M$ and $w_c \ll N$. Codes with a constant row and column weight are called *regular* codes. *Irregular* codes allow different row and column weights in each row and column and usually achieve better BER performance than regular codes [CFRU01]. Irregular codes can be described by maximum and minimum row/column weights and the distribution of weights. Codes for a desired block length can be computed using a method called density evolution [RSU01].

6.1.2 Representation by Tanner graphs

An LDPC code can be represented by a bipartite graph, whose nodes are divided into two disjoint sets called bit nodes (or variable nodes) and check nodes. Every edge connects a bit node to a check node. Bipartite graphs for block codes have been proposed by Tanner [Tan81] and therefore are denoted as Tanner graphs.

Figure 6.1 shows a Tanner graph for a (7,4)-Hamming code, which is defined by the following parity check matrix:

$$\mathbf{H} = \begin{pmatrix} 1 & 1 & 1 & 0 & 1 & 0 & 0 \\ 1 & 1 & 0 & 1 & 0 & 1 & 0 \\ 1 & 0 & 1 & 1 & 0 & 0 & 1 \end{pmatrix} \tag{6.5}$$

The check nodes represent the rows and the bit nodes the columns of the parity check matrix. Edges between bit nodes and check nodes represent the non-zero entries in the parity check matrix. Hence, a set of bit nodes connected to a check node describes one parity check equation. Decoding of LDPC codes is done by passing messages along the edges of the Tanner graphs (see section 6.2).

6.1.3 Quasi-cyclic LDPC codes

In general, the placement of ones in an LDPC matrix is unstructured, which complicates the efficient design of encoders and decoders (in hardware or on a programmable architecture). Quasi-cyclic

Chapter 6 Decoding of quasi-cyclic low density parity check codes

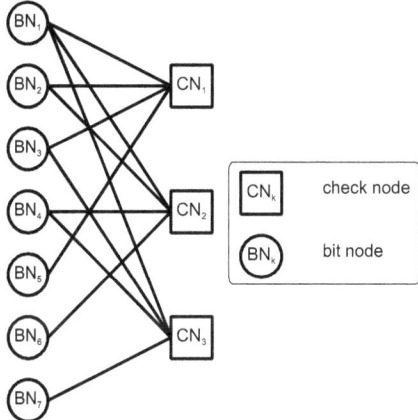

Figure 6.1: Tanner graph of a (7,4) Hamming code

LDPC codes are a class of structured codes [TSS+04, Fos04]; the regular structure simplifies both encoding and decoding.

A quasi-cyclic LDPC matrix is build from structured sub-matrices. A parity check matrix $\mathbf{H}$ of size $M \times N$ is defined by a block matrix $\mathbf{H}_b$ of size $m_b \times n_b$, which describes $z \times z$ sub-matrices of the parity check matrix. The factor z is called the expansion factor. Different codes, with different parity check matrix sizes ($M = m_b \cdot z$, $N = n_b \cdot z$), can be defined by using different expansion factors.

The sub-matrices defined by block matrix $\mathbf{H}_b$ are either empty matrices or circular shifted identity matrices. The entries of $\mathbf{H}_b$ define the circular right shift distance (values between 0 and $z - 1$) and the position of empty sub-matrices (symbolized by $-$). The example below shows the construction of the parity check matrix from a given block matrix $\mathbf{H}_b$ for $z = 3$.[2] Zero elements have been omitted.

$$\mathbf{H}_b = \begin{bmatrix} 0 & - & 2 \\ - & 1 & 1 \end{bmatrix} \tag{6.6}$$

$$\mathbf{H} = \begin{bmatrix} 1 & & & & & & 1 & & \\ & 1 & & & & & & 1 & \\ & & 1 & & & & & & 1 \\ & & & & 1 & & & 1 & \\ & & & & & 1 & & & 1 \\ & & & & 1 & & 1 & & \end{bmatrix} \tag{6.7}$$

The regular structure of quasi-cyclic LDPC codes enables an encoding with linear complexity $\mathcal{O}(N)$ [SYM08]. Furthermore, $z \times z$ sub-matrices can be processed in parallel in the decoder, as indepen-

---
[2] The block matrix $\mathbf{H}_b$ has been arbitrarily chosen, the only purpose is demonstrating the construction of the parity check matrix.

dent pairs of check nodes and bit nodes are processed. Quasi-cyclic LDPC codes are defined in the WiMAX [IEE09b] and IEEE 802.11n [IEE09a] standards.

6.2 Decoding of LDPC codes

The decoding of LDPC codes using Tanner graph is done by iterative message-passing from bit nodes to check nodes and vice versa. One decoding iteration exchanges messages between all bit nodes and all check nodes. The decoding iterates either until all parity check conditions are fulfilled or a maximum number of iterations is reached. Hard-decision decoding is done using the bit flipping algorithm [Gal63]. Below, the focus is on soft-decision decoding. Soft-decision decoders operate on log-likelihood ratios (LLRs), which describe the probability of bit values (see chapter 5.1.3, equation (5.15)). At the end of the LDPC decoding, hard-decision output can be generated by returning the sign bits of the LLR values.

The decoding of LDPC codes by message-passing can be described by the decoding schedule and the applied iterative decoding algorithm.

6.2.1 Decoding schedules

The default schedule for message-passing is the flooding schedule [Gal63, KF98]. The flooding schedule first sends messages from all bit nodes to all check nodes. Then, all check nodes send messages to all bit nodes. Figure 6.2a shows the flooding schedule for the (7,4)-Hamming code from figure 6.1. A flooding schedule allows parallel processing of all nodes. The main drawback of the flooding schedule is its slow convergence rate [SLG07].

Serial message-passing schedules update check nodes and bit nodes in a serial manner, which leads to a faster convergence. A variable node serial schedule processes variable nodes serially. For each variable node v, messages are sent from v to all check nodes connected to v. Next, messages are sent from all check nodes connected to v back to the variable node v. The remaining variable nodes are processed in the same manner.

A check node serial schedule processes check nodes serially. For each check node c, messages are sent from all bit nodes connected to c to the check node. Then, the check node sends back messages to all connected bit nodes. Afterwards, the other check nodes are processed in the same manner. Figure 6.2b shows a check node serial schedule for the (7,4)-Hamming code.

A serial schedule has a faster convergence rate than the flooding schedule, as bit node and check node information is updated during decoding iterations. Sharon, Litsyn and Goldberger [SLG04, SLG07] show that serial schedules converge twice as fast as the flooding schedule. Hence, the number of decoding iterations can be halved and the throughput of a decoder potentially doubles. A serial schedule does not allow parallel processing of all nodes, but the processing may be partially parallelized. In a check node serial schedule, all bit nodes connected to a check node can be processed in parallel. The decoding of quasi-cyclic LDPC codes using a serial schedule can also be done by processing $z \times z$ sub-matrices of the parity check matrix in parallel. Furthermore, serial schedules require less memory than the flooding schedule (section 6.3, [SLG04]).

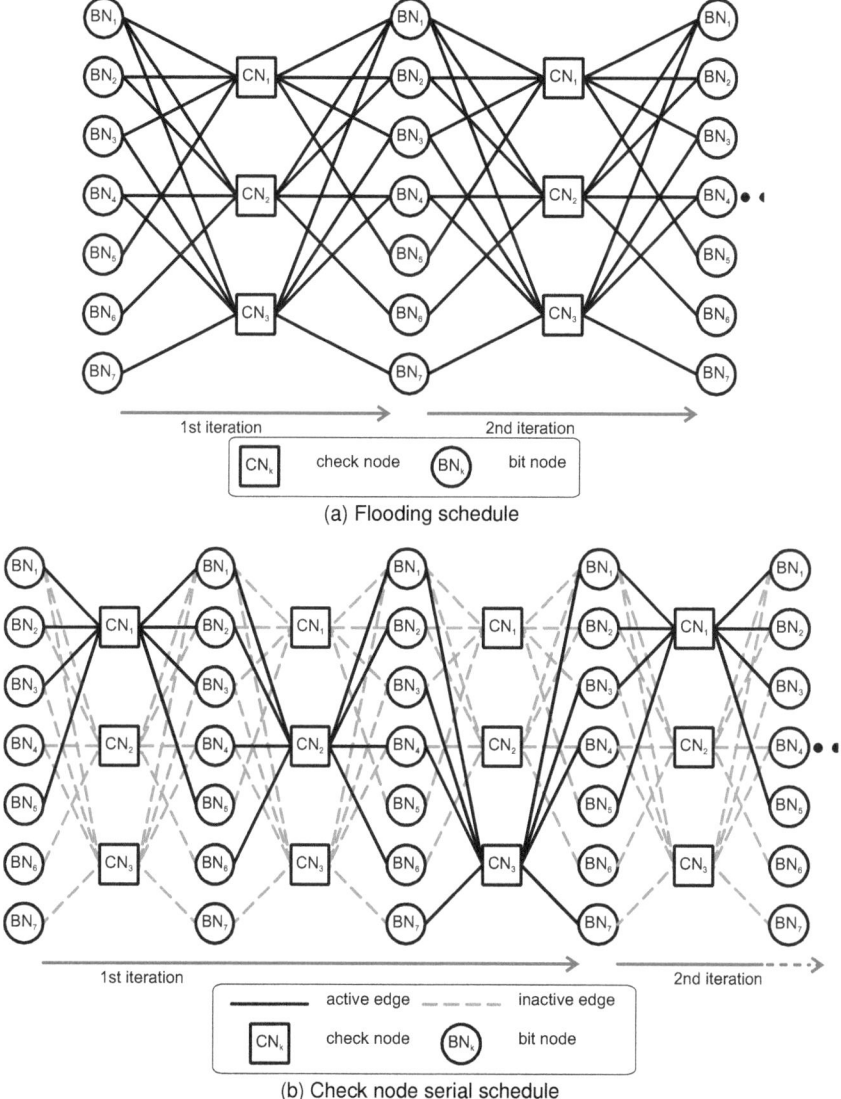

Figure 6.2: LDPC decoding with (a) flooding schedule and (b) check node serial schedule

6.2.2 Iterative decoding algorithms

Decoding algorithms define the messages that are exchanged between check nodes and bit nodes. The input for the decoding are the LLRs obtained from demodulation (I_n). These LLR values define the intrinsic bit node information L_n^i of bit node n in the first decoding iteration. The message from bit node n to check node m is denoted as $L_{n,m}^i$. The state of the check nodes is defined by the extrinsic information $E_{m,n}^i$ for all incoming edges. The set of bit nodes connected to check node m is denoted as $\mathcal{N}(m)$. During the message-passing, the extrinsic check node information for the currently processed edge from bit node n is computed from all other incoming edges, except the edge from node n. The reduced set of bit nodes is denoted as $\mathcal{S}(m,n) = \mathcal{N}(m) \setminus \{n\}$. The notation is summarized in table 6.1.

Table 6.1: Notation for LDPC decoding

Notation	Definition
n	Bit node index
m	Check node index
I_n	Input LLR for bit node n
L_n^i	Intrinsic information of bit node n (in decoding iteration i)
$L_{n,m}^i$	Message from bit node n to check node m
$E_{m,n}^i$	Extrinsic information from check node m to bit node n
$\mathcal{N}(m)$	Set of bit nodes connected to check node m
$\mathcal{M}(n)$	Set of check nodes connected to bit node n
$\mathcal{S}(m,n)$	Set of bit nodes connected to check node m, excluding node n ($\mathcal{N}(m) \setminus \{n\}$)

The most popular algorithms for soft-decision LDPC decoding are the sum-product or belief propagation (BP) algorithm [Gal63, Mac99] and various algorithms based on the min-sum algorithm [Wib96, CDE+05]. Both algorithms can be described by the following three equations:

$$L_{n,m}^i = L_n - E_{m,n}^i \qquad (6.8)$$

$$E_{m,n}^{i+1} = f\left(L_{n',m}^i | n' \in \mathcal{N}(m) \setminus \{n\}\right) \qquad (6.9)$$

$$L_n^{i+1} = L_{n,m}^i + E_{m,n}^{i+1} \qquad (6.10)$$

The message to the check node $L_{n,m}^i$ is calculated by subtracting the old extrinsic check node information for node n from the intrinsic information L_n^i of the bit node, later the new extrinsic check node information is added to $L_{n,m}^i$. Figure 6.3 shows the complete decoding algorithm for a check node serial schedule.

Chapter 6 Decoding of quasi-cyclic low density parity check codes

The min-sum algorithm and the BP algorithm differ in the update function for the check node information, as can be seen in the following equations:

$$f_{\text{BP}}\left(L^i_{n',m}|n' \in \mathcal{S}(m,n)\right) = \prod_{n' \in \mathcal{S}(m,n)} \text{sign}\left(L^i_{n',m}\right) \cdot \Psi\left(\sum_{n' \in \mathcal{S}(m,n)} \Psi\left(L^i_{n',m}\right)\right) \quad (6.11)$$

with $\Psi(x) = -\log_2\left(|\tanh(x/2)|\right)$

$$f_{\text{min-sum}}\left(L^i_{n',m}|n' \in \mathcal{S}(m,n)\right) = \prod_{n' \in \mathcal{S}(m,n)} \text{sign}\left(L^i_{n',m}\right) \cdot \min_{n' \in \mathcal{S}(m,n)} \left|L^i_{n',m}\right| \quad (6.12)$$

```
1:  Initialization
2:  i ← 0
3:  E⁰_{m,n} ← 0, L⁰_n ← I_n    ∀n ∈ [1, N], ∀m ∈ M(n)
4:  Decoding iteration
5:  while i ≤ i_max and parity check fails do
6:    Loop on check nodes
7:    for s = 1 to M do
8:      Messages to check node
9:      for all n ∈ N(m) do
10:       L^i_{n,m} ← L^i_n − E^i_{m,n}
11:     end for
12:     Messages to bit nodes
13:     for all n ∈ N(m) do
14:       E^{i+1}_{n,m} ← f(L^i_{n',m}|n' ∈ S(m,n))
15:       L^{i+1}_n ← L^i_{n,m} + E^{i+1}_{n,m}
16:     end for
17:   end for
18:   i ← i + 1
19: end while
```

Figure 6.3: Check node serial iterative decoding algorithm for $M \times N$ LDPC codes

The BP algorithm (equation (6.11)) has a significantly higher computational complexity than the min-sum algorithm (equation (6.12)). Furthermore, the min-sum algorithm requires much less memory [ZC09]. Therefore, the min-sum algorithm has been chosen for the LDPC decoder implementation on the scalable SIMD processor architecture.

The memory requirements of the min-sum algorithm are low, as it only requires the sign bits of messages from bit nodes and the minimum of all messages $n' \in \mathcal{N}(m) \setminus \{n\}$. The minimum is either the absolute minimum or the second smallest value (the second minimum), depending on the value of n. Hence, it is sufficient to store these two minima, the position of the absolute minimum (for comparing with the node position n), and the sign bits of all messages.

6.3 SIMD implementation of LDPC decoding for WiMAX

Decoding algorithms for quasi-cyclic LDPC codes defined in the WiMAX standard [IEE09b] have been implemented on the proposed SIMD processor architecture to analyze the scalability of SIMD processing for LDPC decoding. WiMAX defines multiple codes with different expansion factors z from the same block matrices $\mathbf{H}_b$. The values of the expansion factor range from 24 to 96 with a stride of four. Codes for z factors 32, 64, and 96 and code rates $R = 2/3$ and $R = 5/6$ have been implemented. Code parameters are listed in table 6.2, the underlying block matrices are shown in table 6.3, empty sub-matrices are symbolized by gray boxes. The block matrices $\mathbf{H}_b$ define cyclic shift distances for expansion factor $z = 96$, the shift distance for other values of the expansion factor z is calculated as:

$$[\mathbf{H}_{b,z}](i,j) = \left\lfloor \frac{[\mathbf{H}_b](i,j) \cdot z}{96} \right\rfloor \qquad (6.13)$$

Table 6.2: Implemented WiMAX codes [IEE09b]

N [bit]	z factor	K [bit]		Number of slots		
		$R = 2/3$	$R = 5/6$	QPSK	16-QAM	64-QAM
768	32	512	640	8	4	—
1536	64	1024	1280	16	8	—
2304	96	1536	1920	24	12	8

Table 6.3: Block matrices for WiMAX [IEE09b]

$\mathbf{H}_b$ for code rate $R = 5/6$

1	25	55		47	4		91	84	8	86	52	82	33	5	0	36	20	4	77	80	0		
	6		36	40	47	12	79	47		41	21	12	71	14	72	0		44	49	0	0	0	0
51	81	83	4	67		21		31	24	91	61	81	9	86	78	60	88	67	15			0	0
68		50	15		36	13	10	11	20	53	90	29	92	57	30	84	92	11	66	80			0

$\mathbf{H}_b$ for code rate $R = 2/3$

3	0		2	0		3	7		1	1				1	0								
		1		36		34	10			18	2		3	0		0	0						
		12	2		15	40		3		15		2	13				0	0					
			19	24		3	0		6		17			8	39			0	0				
20		6			10	29		28			14	38				0			0	0			
		10		28	20			8		36		9		21	45					0	0		
35	25		37		21			5			0		4	20							0	0	
	6	6				4		14	30		3	36		14		1							0

Chapter 6 Decoding of quasi-cyclic low density parity check codes

Different z factors and code rates have been selected to analyze the influence of changing parameters. The codes with expansion factor $z = 96$ are the computationally most challenging codes, as they process the biggest matrices and require the most memory for storing check node information. The implemented LDPC decoders use 16-bit vector elements for the LLRs, as the scalable SIMD architecture only supports 16-bit operations. Hardware implementations of LDPC decoders (e.g. [MS03, RdBKC06, SC08, KW08, MB06]) usually use word lengths between four and eight bits for LLRs. Hence, one way to achieve better throughput is implementing support of shorter vector element types on the scalable SIMD architecture.

The implemented LDPC decoders apply a check node serial schedule and use the min-sum algorithm for decoding. Due to the quasi-cyclic structure of LDPC codes for WiMAX, $z \times z$ sub-matrices of the parity check matrix can be processed in parallel (i.e. z bit nodes and z check nodes). The serial schedule also allows processing multiple sub-matrices in a row of $\mathbf{H}_b$ in parallel, as all corresponding bit nodes are connected to the same z check nodes. For SIMD widths less than 1024 bit and SIMD width 1024 bit with z factor 64, the parallel processing of one sub-matrix is sufficient, as the expansion factor z is a multiple of the number of 16-bit elements in a vector. The remaining codes require the parallel processing of multiple $z \times z$ sub-matrices on a 1024-bit SIMD processor architecture to utilize all SIMD lanes efficiently.

Below, the basic algorithm for min-sum decoding and its memory requirements are explained first. Next, the algorithm implementation for the parallel processing of one sub-matrix is presented. In the last subsection, the extension to the parallel processing of multiple sub-matrices and the limitations of this approach are discussed. The implemented LDPC decoders perform a fixed number of iterations and generate hard-decision output afterwards.

6.3.1 Algorithm for min-sum decoding

The min-sum decoding algorithm based on equations (6.8), (6.10), and (6.12) consists of two processing steps: calculating messages and updating the check node state based on incoming messages. The check node state can be described by the parameters in table 6.4. The algorithms for calculating messages and updating the check node state are shown in figure 6.4 and figure 6.5 respectively.

Table 6.4: Notation for min-sum decoding

Notation	Definition
$p^i_{\text{sign},m}$	Product of sign bits of incoming messages at check node m
$M^i_{m,\text{min}}$	Minimum of incoming messages at check node m
$M^i_{m,\text{min2}}$	Second minimum (second smallest value) of incoming messages at check node m
$M^i_{m,\text{pos}}$	Bit node position of the minimum of incoming messages at check node m
$s^i_{n,m}$	Sign bit for message form bit node n to check node m

Chapter 6 Decoding of quasi-cyclic low density parity check codes

1: **if** $n \neq M^i_{m,\text{pos}}$ **then**
2: $\quad |E^i_{m,n}| \leftarrow M^i_{m,\text{min}}$
3: **else**
4: $\quad |E^i_{m,n}| \leftarrow M^i_{m,\text{min2}}$
5: **end if**
6: $s \leftarrow p^i_{\text{sign},m} \oplus s^i_{n,m}$
7: $L^i_{n,m} \leftarrow L^i_n - (-1)^s \cdot |E^i_{m,n}|$

Figure 6.4: Message calculation for message from bit node n to check node m using the min-sum algorithm

During the calculation of message $L^i_{n,m}$ from bit node to check node, the sign of $E^i_{m,n}$ is calculated by an exclusive or of the sign $s^i_{n,m}$ for the message from the current bit node to the current check node and the sign product of all messages from bit nodes $p^i_{\text{sign},m}$. The amplitude is selected based on the position of the current node n and the minimum position $M^i_{m,\text{pos}}$. Based on the calculated sign s, $|E^i_{m,n}|$ has either to be subtracted or added to the intrinsic information L^i_n of the bit node. This can be implemented by a conditional add/subtract operation, using the sign bit as condition. The update of the bit node based on the extrinsic information of the check node can be calculated in a similar manner.

1: $M^{i+1}_{m,\text{min}} \leftarrow \text{MAXINT}$
2: $M^{i+1}_{m,\text{min2}} \leftarrow \text{MAXINT}$
3: $p^{i+1}_{\text{sign},m} \leftarrow 0$
4: **for all** $n \in \mathcal{N}(m)$ **do**
5: $\quad s^{i+1}_{n,m} \leftarrow \text{sign}\left(L^i_{n,m}\right)$
6: $\quad p^{i+1}_{\text{sign},m} \leftarrow p^{i+1}_{\text{sign},m} \oplus s^{i+1}_{n,m}$
7: $\quad$ **if** $|L^i_{n,m}| < M^{i+1}_{m,\text{min}}$ **then**
8: $\quad\quad tmp \leftarrow M^{i+1}_{m,\text{min}}$
9: $\quad\quad M^{i+1}_{m,\text{min}} \leftarrow |L^i_{n,m}|$
10: $\quad\quad M^{i+1}_{m,\text{pos}} \leftarrow n$
11: $\quad$ **else**
12: $\quad\quad tmp \leftarrow |L^i_{n,m}|$
13: $\quad$ **end if**
14: $\quad$ **if** $tmp < M^{i+1}_{m,\text{min2}}$ **then**
15: $\quad\quad M^{i+1}_{m,\text{min2}} \leftarrow tmp$
16: $\quad$ **end if**
17: **end for**

Figure 6.5: Check node update based on incoming message for min-sum algorithm

The update of the check node state (figure 6.5), consists of updating the minimum, second minimum (the second smallest value), minimum position, and sign product. The sign bit of the incoming message also has to be saved.

The memory requirements of the min-sum algorithm are low. The signs $s_{n,m}^i$ and the sign product $p_{\text{sign},m}^i$ can be computed using mask vectors and later merged in 16-bit values to reduce the memory requirements. Overall, the min-sum algorithm requires saving minima ($M_{m,\text{min}}^i$, $M_{m,\text{min2}}^i$), minimum position $M_{m,\text{pos}}^i$, signs, and the messages from the bit nodes ($L_{n,m}^i$), which are needed for the following update of the bit nodes. If a serial schedule is applied to the decoding process, the intrinsic bit node information L_n^i can be overwritten with the message to the check node $L_{n,m}^i$, which further reduces the memory requirements. For N bit nodes, M check nodes and a regular code with constant row weight w_r the total memory requirement in 16-bit values is:

$$MEM = N + M \cdot \left(3 + \left\lceil \frac{w_r + 1}{16} \right\rceil \right) \tag{6.14}$$

6.3.2 Implementation for the parallel processing of one sub-matrix

The min-sum algorithm can be applied to $z \times z$ sub-matrices in parallel for quasi-cyclic LDPC codes. The cyclic shift can be implemented by one or several vector rotation operations depending on the number of vectors required for z bit nodes. If multiple vectors are required, data from consecutive vectors has to be merged by masked move operations. An example for $z = 12$ and $V = 4$ elements in a vector is shown in figure 6.6. The vector masks for moving data can be computed from the rotation distance using a special instruction, which sets the first k mask elements to one based on a scalar input k.

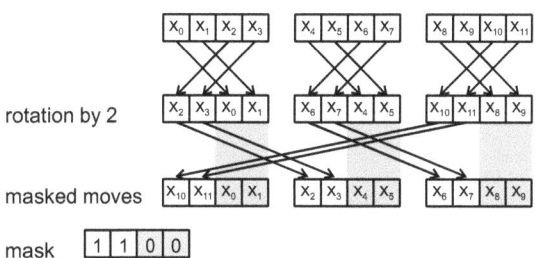

Figure 6.6: Cyclic right shift by two for $z = 12$ and $V = 4$: First, all vectors are rotated, next values from distinct vectors are merged by masked move operations.

The rotation operations enclose the processing of messages from bit nodes to check nodes and, later on, from check nodes back to bit nodes: Before calculating the messages from bit nodes to check nodes for one $z \times z$ sub-matrix, the intrinsic bit node data vectors are right shifted. After finishing the calculations for the bit node updates with the new extrinsic check node information, the shift operation is reversed by a rotation to the left.

The check node or bit node update and cyclic shift or inverse cyclic shift are performed in one loop or split into a pair of loops dependent on the number of data vectors required for the processing of a $z \times z$ sub-matrix. For each vector with intrinsic bit node information L_n^i, three vectors for the old extrinsic check node information $E_{m,n}^i$ (minimum, second minimum, and minimum position) and three vectors for the new extrinsic check node information $E_{m,n}^{i+1}$ are required. All these values have to be stored in general-purpose vector registers (signs are stored in mask registers). As only 16 general-purpose vector registers are available, at most two bit node vectors can be processed in parallel. Yet, the cyclic shift operation requires all bit node vectors. Therefore, the check node or bit node update has to be split into two loops if the expansion factor z is greater than twice the vector length: One loop performs the cyclic shift of z bit nodes; the other loop performs the check node or bit node update. Table 6.5 lists combinations of SIMD width and expansion factor that require separate loops.

Table 6.5: Decomposition of check node/bit node update operations and cyclic shifts into loops: Table entries marked by ✓ do not require separate permutation loops.

SIMD width	$z = 32$	$z = 64$	$z = 96$
128-bit	separate	separate	separate
256-bit	✓	separate	separate
512-bit	✓	✓	separate
1024-bit	✓	✓	separate

After finishing a fixed number of decoding iterations, hard-decision output is generated from the LLR values by selecting the sign bits of LLR values as the hard bits.

6.3.3 Implementation for the parallel processing of multiple sub-matrices

A parallel processing of multiple sub-matrices is necessary if the expansion factor is not a multiple of the vector length ($z = 32$ and $z = 96$, SIMD width 1024 bit, $V = 64$ elements in a vector). Parallel processing of multiple sub-matrices can be done either by processing multiple sub-matrices in one row of the block matrix $\mathbf{H}_b$ in parallel or by processing multiple code words in parallel.

Parallel processing of sub-matrices in a row of $\mathbf{H}_b$

Parallel processing of multiple sub-matrices in one row of $\mathbf{H}_b$ is possible, because the same z check nodes are processed for different bit nodes. Yet, this approach has several disadvantages. Firstly, the parallel processing of sub-matrices requires vector rotation operations on segments of vectors, as different cyclic shifts are necessary for the various sub-matrices. An example is shown in figure 6.7 on the left-hand side. Rotation operations on sub-matrices can be performed on any of the implemented permutation networks; however, these permutations are not supported by rotation instructions. Hence, the permutation patterns have to be manually defined and stored in permutation registers. This approach is laborious and requires a lot of memory for storing the permutation patterns. Secondly, empty sub-matrices require merging data from several data vectors, which introduces further overhead. An example is shown on the right-hand side of figure 6.7. The third disadvantage is that, after processing all non-empty sub-matrices in a row of $\mathbf{H}_b$ during the message-passing from bit nodes to check nodes, the check node information from different vector segments has to be com-

Chapter 6 Decoding of quasi-cyclic low density parity check codes

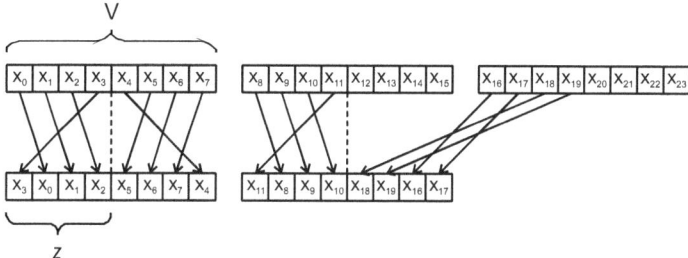

Figure 6.7: Parallel processing of sub-matrices in a row of $\mathbf{H}_b$ for $z = 4$, $V = 8$: The values of the elements of $\mathbf{H}_b$ in the example are $1, 3, 1, -, 2, -$.

bined, because all segments contain information for the same z bit nodes. The combining of check node information from different vector segments leads to a further performance degradation.

Parallel processing of code words

The parallel processing of sub-matrices from different code words is done by merging input data from different code words before the LDPC decoding, e. g. half of the elements in a vector are taken from the first code word and half of the elements are taken from the second code word. Compared to the parallel processing of sub-matrices in a row of $\mathbf{H}_b$, this approach requires more memory, for storing the check node information of two code words, and has a higher initial latency. Yet, an overhead due to empty sub-matrices as well as the combining of check node information from different vector segments can be avoided. The parallel processing of code words also requires rotations on segments of a vector, but the rotation distance is the same for all segments. Dependent on the permutation network, the rotations still can be realized by (modified) vector rotation operations. The parallel processing of code words introduces an overhead for merging and later separating data from different code words. The merging and separating have to be done once per pair of code words; hence, the overhead does not depend on the number of decoding iterations. Furthermore, the separation of code words can be done in one step with the generation of hard-decision output. Overall, the merging and separation of code words increases the runtime by two clock cycles per input vector, as the results in section 6.4 will show, this overhead is negligible.

As the parallel processing of code words has some advantages compared to the parallel processing of sub-matrices in a row of the block matrix $\mathbf{H}_b$, this approach has been selected for the LDPC decoders for expansion factors 32 and 96 on the 1024-bit SIMD processors. Different implementations have been chosen for crossbar and inverse butterfly permutation networks.

On a crossbar network, the merging of code words can be done by interleaving values from code words. The basic approach is shown in figure 6.8a. An interleaving of values enables to perform cyclic shifts by simply doubling the rotation distance. No further changes are necessary for any part of the processing.

Chapter 6 Decoding of quasi-cyclic low density parity check codes

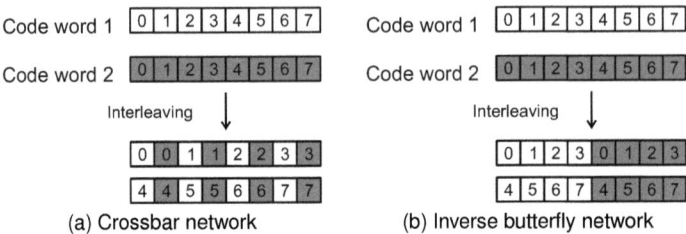

(a) Crossbar network   (b) Inverse butterfly network

Figure 6.8: Merging of code words for parallel processing

On an inverse butterfly network, the interleaving of code words is not a viable approach, as the required permutation is too complex and requires $\log_2(V+1)$ permutation stages (on pairs of vectors). Therefore, a block interleaving (see figure 6.8b) has instead been implemented for the processing on inverse butterfly networks. In a block-interleaved format, cyclic shifts have to be performed by rotations on vector segments. The permutations are defined by permutation patterns in permutation registers. For the LDPC codes with expansion factor $z = 96$, vector masks for the merging of rotated values from different values also have to be stored in memory, as the masks cannot be automatically generated. Figure 6.9 illustrates the required operations for a cyclic shift for an example with $z = 12$ and $V = 8$. The required memory for storing permutation patterns and vector masks is listed in table 6.6.

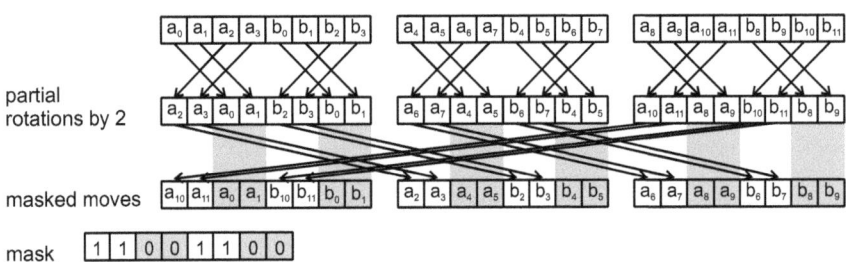

Figure 6.9: Cyclic right shift by two for $z = 12$ and $V = 8$ for block-interleaved parallel processing of code words

Table 6.6: Required memory for permutation patterns and masks for cyclic shifts on segments of vectors on a 1024-bit SIMD processor with an inverse butterfly network

z factor	No. of patterns	Memory
32	80	5 kB
96	80	5 kB + 1 kB for masks

6.4 Performance analysis

In the following, the performance of the decoding of quasi-cyclic LDPC codes for WiMAX on the scalable SIMD processor architecture is analyzed. First, throughput and speedup results are presented. Then, the LIW resource utilization and the performance of bit node and check node update loops are examined. The section concludes with a comparison of the results to other LDPC decoder implementations for WiMAX and an overview of possible application-specific instruction set extensions, which could improve the performance of LDPC decoding.

6.4.1 Throughput and speedup results

Throughput results for the implemented LDPC decoders have been obtained under the assumption that 10 decoding iterations are done, which is an appropriate value for decoding with a serial schedule[3]. The throughput results only contain the user data bits, as parity check bits are no longer needed after the LDPC decoding. The number of user data bits is given by the code rate R and the number of columns in the parity check matrix.

The obtained throughput results are listed in table 6.7. The maximum throughput that can be achieved is 36.58 Mbps for $R = 5/6$ and $z = 64$ on a 1024-bit SIMD processor. The throughput is sufficient for 3G systems, yet nowhere close to the requirements of 4G systems. The throughput for $R = 2/3$ is lower as the number of user data bits is reduced and as a different LDPC matrix is used. The LDPC matrices for $R = 2/3$ have twice as many rows and half as many ones per row as the LDPC matrices for $R = 5/6$. Hence, the overall complexity is the same, except for the initialization of check nodes (i. e. $M^i_{m,\min}$, $M^i_{m,\min 2}$), which requires twice as many operations for $R = 2/3$.

Variations in throughput results for different expansion factors occur due to different implementations of cyclic shift operations, i e. if the cyclic shift operations have to be done in a separate loop from the check node/bit node processing, the performance is reduced.

Table 6.7: Throughput of LDPC decoding with 10 decoding iterations

SIMD architecture	Throughput [Mbps], $R = 5/6$			Throughput [Mbps], $R = 2/3$		
	$z = 32$	$z = 64$	$z = 96$	$z = 32$	$z = 64$	$z = 96$
128-bit	3.93	4.05	3.94	3.11	3.20	3.04
256-bit Bfy1/Cross1	8.84	7.87	7.88	6.82	6.21	5.87
256-bit Bfy2/Cross2	9.16	7.87	7.88	7.04	6.21	5.87
512-bit Bfy1/Cross1	18.29	17.85	15.12	14.04	13.88	11.63
512-bit Bfy2/Cross2	18.29	18.50	15.12	14.04	14.35	11.63
1024-bit Bfy1	36.42	36.59	29.39	27.95	28.07	22.62
1024-bit Cross1	36.41	36.59	30.13	27.94	28.07	23.17
1024-bit Bfy2	36.28	36.59	29.39	27.75	28.07	22.62
1024-bit Cross2	36.41	36.59	30.13	27.94	28.07	23.17

---

[3]The referenced architectures in section 6.4.3 also use at most 10 decoding iterations.

Speedup results for the LDPC decoding are presented in figure 6.10. The speedup is measured in comparison to the performance on a 128-bit SIMD processor (all 128-bit SIMD processors achieve the same throughput). The speedup results show minimal differences between codes with rate $R = 5/6$ and codes with rate $R = 2/3$.

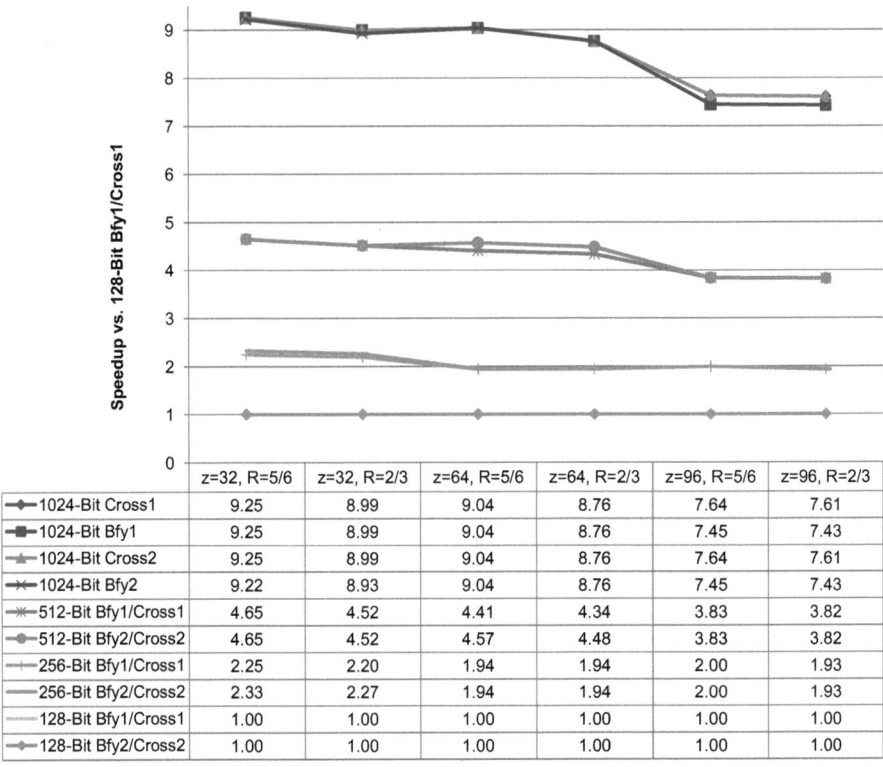

	z=32, R=5/6	z=32, R=2/3	z=64, R=5/6	z=64, R=2/3	z=96, R=5/6	z=96, R=2/3
1024-Bit Cross1	9.25	8.99	9.04	8.76	7.64	7.61
1024-Bit Bfy1	9.25	8.99	9.04	8.76	7.45	7.43
1024-Bit Cross2	9.25	8.99	9.04	8.76	7.64	7.61
1024-Bit Bfy2	9.22	8.93	9.04	8.76	7.45	7.43
512-Bit Bfy1/Cross1	4.65	4.52	4.41	4.34	3.83	3.82
512-Bit Bfy2/Cross2	4.65	4.52	4.57	4.48	3.83	3.82
256-Bit Bfy1/Cross1	2.25	2.20	1.94	1.94	2.00	1.93
256-Bit Bfy2/Cross2	2.33	2.27	1.94	1.94	2.00	1.93
128-Bit Bfy1/Cross1	1.00	1.00	1.00	1.00	1.00	1.00
128-Bit Bfy2/Cross2	1.00	1.00	1.00	1.00	1.00	1.00

Figure 6.10: Speedup of LDPC decoding compared to a 128-bit SIMD processor with a single-vector inverse butterfly network

Speedups better than linear speedup can be achieved if an increasing SIMD width leads to a more efficient decomposition of the processing stages. For example, the LDPC decoder implementation for $z = 32$ on a 128-bit SIMD processor architecture requires separate loops for the cyclic shift permutations (see table 6.5). The pair of loops for the cyclic shift and the inverse cyclic shift requires four clock cycles per vector for loading and storing data vectors, which leads to a decreased performance. If the SIMD width is increased from 128 to 256 bit, the permutations can be combined with the check node and the bit node update, leading to a better performance.

On a 1024-bit SIMD processor architecture, the decoding for expansion factors $z = 32$ and $z = 96$ is done by processing sub-matrices from two code words in parallel, because the expansion factors are no longer multiples of the SIMD vector length. The speedup results for parallel processing of code words are similar to the speedup results for the parallel processing of one $z \times z$ sub-matrix. Hence, the expansion factor z is no hard limit for parallel SIMD processing. For $z = 96$, the speedup on inverse butterfly networks is degraded (for 1024-bit SIMD processors), because vector masks and permutation patterns for the cyclic shift operation on segments of vectors have to be read from memory. The memory access cannot be hidden by LIW processing, as the performance of the loops that realize the cyclic shift operations is already determined by memory access. For $z = 32$, permutation patterns can be loaded in parallel to the processing of messages from bit node to check node or vice versa — vector masks are not required at all for $z = 32$.

6.4.2 LIW resource utilization

Table 6.8 describes resource utilization and throughput of check node and bit node update loops for all SIMD processors and all implemented expansion factors. The performance of permutation loops is not discussed. Separate permutation loops for cyclic shifting of z bit nodes always require two clock cycles per vector for loading and storing the bit nodes. The permutations can be done in parallel to memory access.

In table 6.8, the column marked with *Rot.* distinguishes loops with combined bit/check node update and cyclic shift (marked by ✓) from loops without cyclic shift (i. e. the shifting is done in separate loops). The table lists the average number of parallel operations in one instruction ($N_{\text{par.}\varnothing}$) and the resource utilization of the VALU (R_{VALU}). The resource utilization of the VMAC is not listed, as the LDPC decoding does not require MAC operations or multiplications. The last column lists the processing time per bit node vector in clock cycles.

A check node update loop without permutations requires eight clock cycles per vector for calculating the messages from bit nodes to check nodes and updating check node minima and sign bits. The resource utilization of the VALU is 100 percent; hence, a faster processing is not possible. A bit node update loop without permutations takes three clock cycles per vector for calculating the messages from check nodes to bit nodes and updating the bit nodes. Again, the VALU is performing useful computations in each clock cycle. Furthermore, the average number of parallel operations per LIW instructions is close to the number of available instruction slots. In sum, the check node and bit node updates require 11 clock cycles per vector if the cyclic shift of bit nodes is done in a separate loop.

In case permutations for cyclic shifts are performed in the bit node and check node update loops, the processing of both loops requires on average between 12.5 and 13 clock cycles per vector, which is an increase of between 1.5 and 2 clock cycles. As performing cyclic shifts in separate loops requires 4 clock cycles per vector, the throughput is higher if cyclic shifts can be combined with the check node and the bit node update loops.

Table 6.8: LIW resource utilization of bit node and check node update loops

SIMD width	z	Description	Rot.	$N_{par.\varnothing}$	R_{VALU}	Cycles/ vector
256-bit Bfy1/Cross1	32	Check node update	✓	2.667	88.89 %	9
		Bit node update		3.875	75 %	4
256-bit Bfy2/Cross2		Check node update	✓	2.588	94.12 %	8.5
		Bit node update		3.375	75 %	4
512-bit Bfy1/Cross1	32	Check node update	✓	2.875	100 %	8
		Bit node update		3.333	66.67 %	4.5
512-bit Bfy2/Cross2		Check node update	✓	3.000	100 %	8
		Bit node update		3.556	66.67 %	4.5
512-bit Bfy1/Cross1	64	Check node update	✓	2.667	88.89 %	9
		Bit node update		3.875	75 %	4
512-bit Bfy2/Cross2		Check node update	✓	2.588	94.12 %	8.5
		Bit node update		3.375	75 %	4
1024-bit Bfy1	32	Check node update	✓	2.706	94.12 %	8.5
		Bit node update		3.5	75 %	4
1024-bit Cross1		Check node update	✓	2.875	100 %	8
		Bit node update		3.33	66.67 %	4.5
1024-bit Bfy2		Check node update	✓	2.647	94.12 %	8.5
		Bit node update		3.375	75 %	4
1024-bit Cross2		Check node update	✓	3.000	100 %	8
		Bit node update		3.56	66.67 %	4.5
1024-bit Bfy1/Cross1	64	Check node update	✓	2.875	100 %	8
		Bit node update		3.333	66.67 %	4.5
1024-bit Bfy2/Cross2		Check node update	✓	3.000	100 %	8
		Bit node update		3.556	66.67 %	4.5
Other combinations of z and SIMD width		Check node update		2.625	100 %	8
		Bit node update		3.833	100 %	3

6.4.3 Comparison to other architectures

Table 6.9 lists the performance and parameters of selected ASIC implementations and one SDR implementation of LDPC decoding for WiMAX. The ASIC implementations with the highest throughputs and the greatest flexibility have been listed.

The LDPC decoders implemented by Kuo and Willson [KW08] and by Sun and Cavallaro [SC08] support all WiMAX code rates and expansion factors. The decoder in [SC08] also supports all quasi-cyclic LDPC codes defined by IEEE 802.11n. The LDPC decoder chip by Shih et al. [SZLW07] supports all rate 1/2 codes defined by WiMAX.

The decoder by Sun and Cavallaro achieves the highest throughput of all ASIC implementations with a maximum throughput of 1 Gbps, at the cost of high power consumption. The decoder by Shih et al. achieves the lowest power consumption — and the lowest throughput.

In comparison to these architectures, the throughput results on the scalable SIMD processor architecture are close to the throughput results on the low-power LDPC decoder chip by Shih et al. (max. 36.59 Mbps on a 1024-bit SIMD processor for $z = 64$ and $R = 5/6$), while the power consumption is closer to the high-performance LDPC decoder by Sun and Cavallaro (≈ 287 mW on a 1024-bit SIMD

Table 6.9: Overview of LDPC decoder implementations for WiMAX

	[KW08] ASIC	[SC08] ASIC	[SZLW07] ASIC	[SMZC07] SODA PE
Technology	0.18 µm	90 nm	0.13 µm	0.18 µm
Parallel units	24	96	—	32
Frequency	100 MHz	400 MHz	83.3 MHz	400 MHz
Code rate	all	all	1/2	5/6
z factor	all	all	all	96
Quantization	5 bit	8 bit	—	16 bit
Max. iterations	10	10	8	10
Algorithm	min-sum	BP	min-sum	min-sum
Max. throughput	68 Mbps	1 Gbps	30.3 Mbps	18.3 Mbps / 30.4 Mbps (ASP)
Area	55 kgates	$13.5\,\text{mm}^2$	$8.29\,\text{mm}^2$	$\approx 5.1\,\text{mm}^2$
Power [mW]	165	410	52	≈ 730

processor with a single-vector permutation network for $z = 64$ and $R = 5/6$ without memories, see chapter 7.1). One reason for the low throughput on the scalable SIMD processor architecture is the used quantization of LLRs. As the SIMD processor architecture only supports 16-bit vector elements, LLRs are represented by 16-bit values. LDPC decoding in ASICs is usually done with LLRs values quantized to between four and eight bits.

Seo et al. [SMZC07] implemented LDPC decoding for one WiMAX code ($z = 96$, $R = 5/6$) on one 512-bit SIMD SODA PE. At the maximum clock frequency of 400 MHz, a decoding throughput of 18.3 Mbps is achieved, which is approximately 21 percent higher than the throughput on one of the proposed 512-bit SIMD processors at 300 MHz. Seo et al. also presented an LDPC decoding ASP based on SODA. The ASP contains LDPC accelerator units, which perform the processing of the extrinsic check node information $E^i_{m,n}$. Furthermore, memory units perform the cyclic shifting of $z \times z$ sub-matrices during memory access. Hence, there is no need for vector permutations. As a third optimization, buffers have been designed, which store messages between check nodes and bit nodes as well as the extrinsic check node information. Special instructions allow accessing these buffers. The optimizations lead to an increased throughput of 30.4 Mbps.

6.4.4 Improving the LDPC decoding performance

In the following, possible modifications on the scalable SIMD processor architecture that could potentially lead to significant performance gains are discussed. Modifications for improving the SIMD scalability (i. e. for SIMD widths greater than the expansion factor z) are discussed in chapter 7.3 and not considered in this section.

Quantization of LLR values

Assuming that the z factor is sufficiently large compared to the SIMD vector length, the LDPC decoding throughput could be doubled by adding support for 8-bit data types in the VALU and the VPU and performing two 8-bit operations instead of one 16-bit operation. Changes on the other SIMD processing units are not required. 8-bit support increases the complexity of the permutation network (e.g. one additional permutation stage for an inverse butterfly network) and requires some small modifications on the VALU. Furthermore, mask values also need to be provided for each 8-bit vector element.

In principle, the word length of LLR values could be further reduced to four bits, leading to a potential quadrupling of the throughput. However, the quantization leads to a degraded BER performance.

Instruction set extensions for LDPC decoding

The performance of LDPC decoding on the scalable SIMD architecture is limited by the utilization of the VALU, which is required for computing messages from bit nodes to check nodes (and vice versa) and the update of the extrinsic check node information. The runtime of both processing steps could potentially be reduced by custom instructions.

The message computations can be sped up by a custom instruction for the calculation of $E_{m,n}^i$ (and $E_{m,n}^{i+1}$), which implements the algorithm segment in figure 6.11. The custom instruction selects the absolute value of the extrinsic check node information based on a comparison of the current node position n and the minimum position. The sign bit of $E_{m,n}^i$ is computed by an exclusive or operation of the sign product and the sign bit for the current position.

1: **if** $n \neq M_{m,\text{pos}}^i$ **then**
2: $\quad |E_{m,n}^i| \leftarrow M_{m,\text{min}}^i$
3: **else**
4: $\quad |E_{m,n}^i| \leftarrow M_{m,\text{min2}}^i$
5: **end if**
6: $E_{m,n}^i \leftarrow (-1)^{\left(p_{\text{sign},m}^i \oplus s_{n,m}^i\right)} \cdot |E_{m,n}^i|$

Figure 6.11: Calculation of $E_{m,n}^i$ during the message-passing

The performance of the check node update could be improved by adding a custom instruction for the update of the minimum $M_{m,\text{min}}^i$ and its position based on an incoming message $x = |L_{n,m}^i|$. The custom instruction in figure 6.12 compares and conditionally swaps the incoming message x and the minimum $M_{m,\text{min}}^i$. Furthermore, the minimum position is updated.

Both proposed instructions require additional register ports for accessing all operands. The number of register ports could be reduced by adding local state registers to the processing units, which contain the check node state, i.e. the minima $M_{m,\text{min}}^i$ and $M_{m,\text{min2}}^i$, the minimum position $M_{m,\text{pos}}^i$ and the required sign bits $s_{n,m}^i$ and $p_{\text{sign},m}^i$. In this case, further instructions have to be added for initializing and reading the local registers.

```
1: if $x < M^i_{m,\text{min}}$ then
2:     $x \leftarrow M^i_{m,\text{min}}$
3:     $M^i_{m,\text{min}} \leftarrow x$
4:     $M^i_{m,\text{pos}} \leftarrow n$
5: end if
```

Figure 6.12: Update of minimum and minimum position during the check node update

6.5 Conclusion

Parallel SIMD processing of quasi-cyclic LDPC codes can be efficiently done by processing $z \times z$ submatrices in parallel. Yet, the results in section 6.4 also demonstrate that parallel SIMD processing is not limited by the expansion factor z. If the expansion factor z is not equal to or a multiple of the SIMD width, close to linear speedup can still be achieved by processing multiple code words in parallel. However, this approach requires more complex permutation patterns for cyclic shift operations on vector segments. Hence, the expansion factor z can be seen as a soft limit for the scalability of the SIMD vector length: performance gains are possible for vector lengths greater than z, yet the vector length should not be set to arbitrarily large values.

While LDPC decoding can be done on the proposed scalable SIMD architecture, throughput and power consumption are significantly worse than in state-of-the-art LDPC decoding ASICs. One reason for the low throughput and the high power consumption is the quantization of LLR values (16 bits). A reduced word length of eight bits for LLR values would not lead to a significant performance degradation and enable to either achieve higher throughput (by processing twice as many bit nodes in parallel) or reduce the power consumption (by reducing the SIMD bit width). The throughput is further limited by the instruction set, as message-passing and check node update have to be done by a series of operations on the VALU.

If the flexibility of a programmable processor architecture is desired for LDPC decoding, the proposed SIMD processor architecture should be replaced by an optimized ASIP based on SIMD processing. Such an ASIP could implement custom instructions for LDPC decoding (e. g. as in section 6.4.4) and possibly the decoding of turbo codes as well. Furthermore, unused processing units (e. g. the VMAC) could be removed, reducing the processor area and power consumption (by reducing the number of register file ports and possibly LIW instruction slots).

Chapter 7

Evaluation of the SIMD architecture efficiency

The previous chapters focused on the mapping of key physical layer algorithms on SIMD processors with LIW support and the scalability of the performance of these algorithms. This chapter completes the evaluation of the scalability of SIMD processing by a discussion of its costs: energy consumption and chip area. Power and energy consumption and chip area estimates have been obtained from the gate level SIMD processor models, using the methodology described in chapter 3.4.

Section 7.1 discusses area and power consumption figures for the implemented SIMD processors. Area and power consumption estimates for data and program memory based on the on-line tool CACTI [TMAJ08] are also presented. The following section (section 7.2) focuses on the energy and area efficiency of SIMD processing. Results show that wide SIMD processors achieve better energy efficiency than SIMD processors with a small number of parallel data paths. Section 7.3 addresses possible architectural changes, which could further improve the scalability by overcoming the limitations of SIMD processing for the discussed algorithms. The last section (section 7.4) addresses another cost of SIMD processing — the parallel programming model. In this context, vectorization techniques for compilers are briefly discussed.

7.1 Area and power consumption results

Table 7.1 summarizes area and average power consumption figures for all synthesized SIMD processor architectures. The average power consumption has been computed by averaging the power consumption of all implemented FFT, sphere decoding, and LDPC decoding algorithms. Next to columns listing chip area in square millimeters and power consumption in milliwatts, the table also contains columns with normalized area and power consumption figures. Area and power have been normalized to a 128-bit SIMD processor with a single-vector inverse butterfly network for vector permutations. This processor architecture has the lowest complexity and achieves the worst throughput results for all implemented algorithms.

Normalized power consumption and area have also been plotted in figure 7.1 and figure 7.2, respectively. The diagrams show four different curves for SIMD processors with single-vector inverse butterfly networks, single-vector crossbar networks, double-vector inverse butterfly networks, and double-vector crossbar networks.

Table 7.1: Area and power consumption results for the scalable SIMD processor architecture

SIMD width	Network	Area [mm²]	Norm. area	Power [mW]	Norm. power
128 bit	Bfy1	0.456	1.00	42.14	1.00
	Cross1	0.456	1.00	41.98	1.00
	Bfy2	0.497	1.09	45.17	1.07
	Cross2	0.508	1.11	45.36	1.08
256 bit	Bfy1	0.831	1.82	74.68	1.77
	Cross1	0.842	1.85	75.49	1.79
	Bfy2	0.914	2.00	81.51	1.93
	Cross2	0.968	2.12	81.46	1.93
512 bit	Bfy1	1.575	3.45	141.89	3.37
	Cross1	1.627	3.57	142.38	3.38
	Bfy2	1.768	3.88	158.32	3.76
	Cross2	1.973	4.33	158.41	3.76
1024 bit	Bfy1	3.140	6.88	289.12	6.86
	Cross1	3.334	7.31	291.75	6.92
	Bfy2	3.507	7.69	321.79	7.64
	Cross2	4.370	9.58	325.44	7.72

7.1.1 Average power consumption

The power consumption results show minimal differences between processors with crossbar and inverse butterfly networks if network width and SIMD width are the same. The difference between crossbar network and inverse butterfly network increases with the SIMD width. A 1024-bit SIMD processor with a double-vector crossbar network consumes 3.65 mW more power than the same processor with a double-vector inverse butterfly network. For 1024-bit SIMD processors with single-vector networks the difference in power consumption between crossbar and inverse butterfly networks is 2.63 mW. Processors with double-vector networks require approximately 10 percent more power than processors with single-vector networks. The increased power consumption results from the greater complexity of the permutation network and the increased number of register file ports, which leads to an increased register file complexity.

Figure 7.1 and table 7.1 show that the power consumption does not double with a doubling of the SIMD width. This can be explained by the overhead for the scalar data path and especially for the LIW decoding: A doubling of the SIMD width leads to double the number of SIMD lanes and an increased complexity of the permutation network, yet area and power consumption of the scalar data path and the instruction decoding logic do not scale at all. Hence, wide SIMD processors are more energy and area efficient than SIMD processors that process narrow vectors — assuming the performance of algorithms scales with the SIMD width.

Table 7.1 and figure 7.1 also show a jump in power consumption from 512-bit to 1024-bit SIMD processors. The power consumption increases due to inefficient RTL code generated from LISA for the distribution of control signals from instruction decoder to SIMD processing lanes: Control signals for the processing in the EX pipeline stage, such as opcodes and register addresses, are distributed

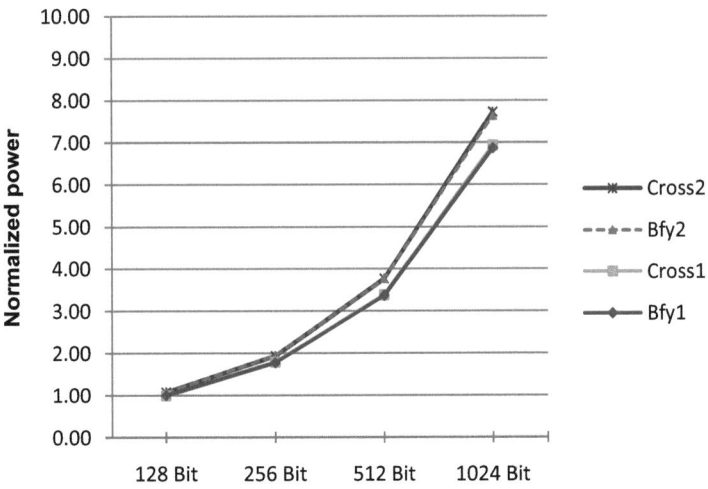

Figure 7.1: Normalized power consumption

to pairs of 16-bit lanes in the RF pipeline stage. Hence, for 64 16-bit lanes a fan-out of 32 is expected. Yet, the RTL code generated by LISA contains enable signals for some of the control signals, which are connected to all signal bit lines. The high fan-out for the enable signals ($32\times$ the number of bits for the control signal) leads to an increased power consumption of the synthesized model. Therefore, wide SIMD processors that are intended as components of actual SDR systems and not just for a design space exploration (as in this paper) should not be modeled in LISA.

7.1.2 Area

The SIMD processor chip area does not double from 128-bit to 256-bit SIMD, because of the constant overhead for the scalar data path and especially the instruction decoding logic (more than 25% of the total area for 128-bit SIMD processors). If the SIMD width is further increased, the relative overhead for instruction decoding and scalar data path is reduced and the slope of the area curves in figure 7.2 is mostly determined by the complexity of the permutation network.

Table 7.2 lists total area and normalized area of the vector permutation unit (VPU) for all implemented permutation networks. Results for the 128-bit SIMD processor with a single-vector crossbar network have not been obtained, because the design hierarchy has been ungrouped during synthesis. The area of the permutation networks depends on the number of 16-bit input elements N and the network topology. Theoretically (see chapter 3.1.5), the area of crossbar networks grows as $\mathcal{O}\left(N^2\right)$, while the area of inverse butterfly networks grows as $\mathcal{O}\left(N\log_2\left(N\right)\right)$. The results in table 7.2 deviate from the expected area growth, because the VPU comprises the permutation network *and* further control logic for selecting operands and permutation patterns as well as logic for bypassing.

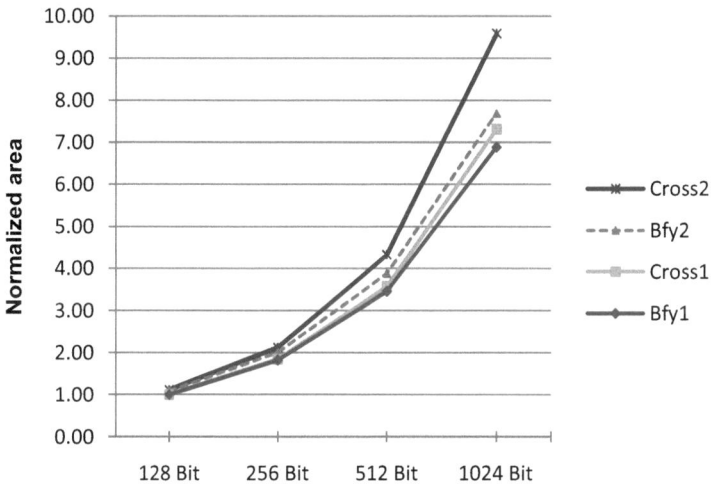

Figure 7.2: Normalized area

7.1.3 Power consumption and area estimates for memories

The results in table 7.1 do not include power consumption and area of program memory and scalar and SIMD data memories as these memories have not been synthesized. Yet, memories contribute to the power consumption and especially the chip area. Therefore, area and power consumption of memories have been estimated using the on-line tool CACTI [TMAJ08], developed by HP Labs. CACTI models the access time, cycle time, area, leakage power, and dynamic power of integrated caches and memories. Different technologies are modeled based on the International Technology Roadmap for Semiconductors (ITRS). Memories have been modeled in 90 nm technology using ITRS low operating power transistor models for a worst-case temperature of 400 Kelvin. The obtained area results are listed in table 7.3, power consumption results are listed in table 7.4. The program memory is modeled as a 24 kB memory with one read port and one write port. The write port is intended for an external control processor or DMA controller, which transfers program code and data to the SIMD processor. The scalar and the SIMD memory are modeled with two combined read/write ports, one for accessing the memory from the SIMD processor, one for an external control processor/DMA controller. The scalar and SIMD memory sizes are 4 kB and 128 kB, respectively.

The results in table 7.3 show that the memories require significantly more area than the processor core. The area of the SIMD memory depends on the SIMD width, as the number of elements in a vector determines the bit width for memory access. Surprisingly, the SIMD memory area decreases from 256-bit to 512-bit. Yet, there is no clear explanation for this behavior in CACTI's area report.

The power consumption figures in table 7.4 contain leakage power and the theoretical maximum dynamic power consumption. The dynamic power consumption assumes one memory access per

Table 7.2: Permutation network area

SIMD width	Network	Area VPU [mm^2]	Norm. VPU area
128 bit	Bfy1	0.009	1.00
	Cross1	—	—
	Bfy2	0.016	1.86
	Cross2	0.022	2.54
256 bit	Bfy1	0.019	2.15
	Cross1	0.024	2.71
	Bfy2	0.034	3.83
	Cross2	0.069	7.80
512 bit	Bfy1	0.039	4.40
	Cross1	0.073	8.25
	Bfy2	0.076	8.64
	Cross2	0.239	27.02
1024 bit	Bfy1	0.087	9.80
	Cross1	0.246	27.82
	Bfy2	0.157	17.80
	Cross2	0.875	98.97

clock cycle (at 300 MHz). The access rates of scalar and SIMD memories are significantly lower for all implemented algorithms; hence, the actual power consumption should be less than the listed power consumption. The power consumption of the SIMD memory increases with the SIMD width, as the port width increases.

As the area and power consumption figures obtained from CACTI's models are only rough estimates, the energy efficiency analysis in the following section is done without considering the power requirements of memories.

7.2 Energy efficiency analysis

Energy consumption is a better processor architecture metric than power consumption, because it takes into account that wider SIMD processors can achieve greater throughputs. The energy consumption of all implemented algorithms on all synthesized SIMD processors has been calculated.

As the energy consumption depends on the power consumption and the algorithm performance, the speedup results obtained in the previous three chapters are summarized in figure 7.3. The implemented algorithms are listed on the abscissa of the diagram: FFT results on the left-hand side, sphere decoding results in the middle, and LDPC decoding results on the right-hand side. The figure shows four groups of curves for 128-bit (speedup approximately one), 256-bit (speedup approximately two), 512-bit (speedup approximately four), and 1024-bit (speedup approximately eight) SIMD processors. The FFT performance has been measured including the overhead for the initialization of parameters.

Chapter 7 Evaluation of the SIMD architecture efficiency

Table 7.3: Area of processor core and memories

SIMD width	Network	Core area	Scalar mem. (4 kB)	Program mem. (24 kB)	SIMD memory (128 kB)	Total area
128 bit	Bfy1	0.46 mm^2	0.13 mm^2	0.98 mm^2	4.56 mm^2	6.13 mm^2
	Cross1	0.46 mm^2	0.13 mm^2	0.98 mm^2	4.56 mm^2	6.13 mm^2
	Bfy2	0.50 mm^2	0.13 mm^2	0.98 mm^2	4.56 mm^2	6.17 mm^2
	Cross2	0.51 mm^2	0.13 mm^2	0.98 mm^2	4.56 mm^2	6.18 mm^2
256 bit	Bfy1	0.83 mm^2	0.13 mm^2	0.98 mm^2	6.85 mm^2	8.79 mm^2
	Cross1	0.84 mm^2	0.13 mm^2	0.98 mm^2	6.85 mm^2	8.80 mm^2
	Bfy2	0.91 mm^2	0.13 mm^2	0.98 mm^2	6.85 mm^2	8.87 mm^2
	Cross2	0.97 mm^2	0.13 mm^2	0.98 mm^2	6.85 mm^2	8.93 mm^2
512 bit	Bfy1	1.57 mm^2	0.13 mm^2	0.98 mm^2	6.56 mm^2	9.24 mm^2
	Cross1	1.63 mm^2	0.13 mm^2	0.98 mm^2	6.56 mm^2	9.30 mm^2
	Bfy2	1.77 mm^2	0.13 mm^2	0.98 mm^2	6.56 mm^2	9.44 mm^2
	Cross2	1.97 mm^2	0.13 mm^2	0.98 mm^2	6.56 mm^2	9.64 mm^2
1024 bit	Bfy1	3.14 mm^2	0.13 mm^2	0.98 mm^2	11.87 mm^2	16.12 mm^2
	Cross1	3.33 mm^2	0.13 mm^2	0.98 mm^2	11.87 mm^2	16.31 mm^2
	Bfy2	3.51 mm^2	0.13 mm^2	0.98 mm^2	11.87 mm^2	16.49 mm^2
	Cross2	4.37 mm^2	0.13 mm^2	0.98 mm^2	11.87 mm^2	17.35 mm^2

Table 7.5 (128-bit, 256-bit SIMD) and table 7.6 (512-bit, 1024-bit SIMD) show the energy consumption results. For radix-2 and mixed-radix FFT implementations, the energy required for processing one FFT has been calculated. The results for hard-decision (FSD) and soft-decision (SFSD) fixed-complexity sphere decoding have been computed for the processing of one OFDM sub-carrier. LDPC decoding energy consumption figures correspond to the decoding of one code word with 10 decoding iterations.

7.2.1 Normalized energy consumption

The results in table 7.5 and table 7.6 show that the total energy consumption usually decreases with an increasing SIMD width. Yet, as the energy consumption also depends on the algorithm complexity, a direct comparison of results is difficult. Therefore, the normalized energy consumption has been calculated as an appropriate indicator for the energy efficiency. The normalized energy consumption is shown in figure 7.4. The figure is split into four parts with normalized energy consumption results for 128-bit, 256-bit, 512-bit, and 1024-bit SIMD processors. The diagrams show the normalized deviation of the energy consumption from the energy consumption of a 128-bit SIMD processor with a single-vector inverse butterfly network, i. e. normalized energy consumption values less than one indicate that less energy is required. Results for different permutation networks for the same SIMD width are grouped together. As in the speedup diagram (figure 7.3), the implemented algorithms are listed on the abscissae of the diagrams. Hence, normalized energy consumption values should be

Table 7.4: Power consumption of processor core and memories

SIMD width	Network	Core power	Scalar mem. (4 kB)	Program mem. (24 kB)	SIMD memory (128 kB)	Total power
128 bit	Bfy1	42.14 mW	2.22 mW	11.95 mW	38.04 mW	94.35 mW
	Cross1	41.98 mW	2.22 mW	11.95 mW	38.04 mW	94.19 mW
	Bfy2	45.17 mW	2.22 mW	11.95 mW	38.04 mW	97.39 mW
	Cross2	45.36 mW	2.22 mW	11.95 mW	38.04 mW	97.58 mW
256 bit	Bfy1	74.68 mW	2.22 mW	11.95 mW	59.77 mW	148.62 mW
	Cross1	75.49 mW	2.22 mW	11.95 mW	59.77 mW	149.43 mW
	Bfy2	81.51 mW	2.22 mW	11.95 mW	59.77 mW	155.45 mW
	Cross2	81.46 mW	2.22 mW	11.95 mW	59.77 mW	155.40 mW
512 bit	Bfy1	141.89 mW	2.22 mW	11.95 mW	76.74 mW	232.80 mW
	Cross1	142.38 mW	2.22 mW	11.95 mW	76.74 mW	233.29 mW
	Bfy2	158.32 mW	2.22 mW	11.95 mW	76.74 mW	249.23 mW
	Cross2	158.41 mW	2.22 mW	11.95 mW	76.74 mW	249.32 mW
1024 bit	Bfy1	289.12 mW	2.22 mW	11.95 mW	131.62 mW	434.91 mW
	Cross1	291.75 mW	2.22 mW	11.95 mW	131.62 mW	437.54 mW
	Bfy2	321.79 mW	2.22 mW	11.95 mW	131.62 mW	467.58 mW
	Cross2	325.44 mW	2.22 mW	11.95 mW	131.62 mW	471.23 mW

compared in the vertical direction of the figure for assessing the impact of a scaling of the SIMD width.

The normalized energy consumption results correlate to the speedup results: If linear speedup or close to linear speedup can be achieved, the energy consumption decreases with an increasing SIMD width. The best energy consumption results have been obtained for the algorithms with the highest speedups, i. e. LDPC decoders for $z = 32$ and $z = 64$ and the 128-point FFT for SIMD widths greater than 256 bits. Algorithms that do not scale as well, e. g. short radix-2 and mixed-radix FFTs, achieve the worst normalized energy consumption figures. In some cases, the energy consumption even increases with an increasing SIMD width.

The normalized energy consumption results for 1024-bit SIMD processors are slightly worse than the results for 512-bit SIMD processors due to the increased power consumption for high fan-out nets (see section 7.1.1). If the LISA generated RTL code would be replaced by hand-optimized RTL code, the energy consumption should steadily decrease with an increasing SIMD width.

A comparison of different permutation networks in figure 7.4 shows that double-vector permutation networks usually require more energy than single-vector networks. The increased energy consumption results mostly from the increased register file complexity, as double-vector networks require one additional read port and one additional write port for the general-purpose SIMD register file. Across all SIMD widths, the power consumption of the register files for double-vector networks is on average twice the power consumption for single-vector networks. Double-vector networks only consume less energy than single-vector networks if the wider permutation network leads to better speedups (e. g. for short radix-2 FFTs).

Table 7.5: Total energy consumption for the implemented algorithms on 128-bit and 256-bit SIMD processors.

Algorithm	128-bit SIMD				256-bit SIMD			
	Bfy1	Cross1	Bfy2	Cross2	Bfy1	Cross1	Bfy2	Cross2
8-pt. FFT	1.24 nJ	1.24 nJ	1.23 nJ	1.10 nJ	2.86 nJ	2.89 nJ	2.81 nJ	2.71 nJ
16-pt. FFT	2.51 nJ	2.50 nJ	2.69 nJ	2.69 nJ	5.97 nJ	6.04 nJ	5.97 nJ	6.01 nJ
32-pt. FFT	5.91 nJ	5.89 nJ	6.34 nJ	6.34 nJ				
64-pt. FFT	14.26 nJ	14.21 nJ	14.99 nJ	15.00 nJ	12.94 nJ	13.08 nJ	13.58 nJ	13.58 nJ
128-pt. FFT	37.10 nJ	36.97 nJ	39.18 nJ	39.21 nJ	33.83 nJ	34.20 nJ	35.75 nJ	35.76 nJ
256-pt. FFT	73.93 nJ	73.66 nJ	78.05 nJ	78.12 nJ	67.01 nJ	67.73 nJ	70.87 nJ	70.88 nJ
512-pt. FFT	165.05 nJ	164.44 nJ	174.54 nJ	174.70 nJ	148.61 nJ	150.22 nJ	157.75 nJ	157.79 nJ
1024-pt. FFT	401.62 nJ	400.16 nJ	425.76 nJ	426.16 nJ	359.62 nJ	363.53 nJ	383.70 nJ	383.80 nJ
2048-pt. FFT	802.57 nJ	799.64 nJ	850.80 nJ	851.59 nJ	718.24 nJ	726.05 nJ	766.40 nJ	766.61 nJ
192-pt. FFT	57.84 nJ	57.63 nJ	61.10 nJ	61.16 nJ	52.53 nJ	53.11 nJ	55.62 nJ	55.63 nJ
384-pt. FFT	123.95 nJ	123.50 nJ	131.08 nJ	131.21 nJ	111.79 nJ	113.00 nJ	118.65 nJ	118.68 nJ
576-pt. FFT	226.38 nJ	225.55 nJ	239.99 nJ	240.21 nJ	197.16 nJ	199.31 nJ	210.20 nJ	210.26 nJ
768-pt. FFT	288.48 nJ	287.43 nJ	305.67 nJ	305.95 nJ	254.42 nJ	257.19 nJ	271.06 nJ	271.14 nJ
960-pt. FFT	394.66 nJ	393.22 nJ	418.60 nJ	418.99 nJ	353.44 nJ	357.28 nJ	377.50 nJ	377.60 nJ
1152-pt. FFT	465.33 nJ	463.63 nJ	493.46 nJ	493.99 nJ	416.63 nJ	421.16 nJ	444.83 nJ	444.95 nJ
4×4 FSD, QPSK	18.66 nJ	18.60 nJ	20.01 nJ	20.18 nJ	16.50 nJ	16.68 nJ	18.00 nJ	17.97 nJ
4×4 FSD, 16-QAM	31.89 nJ	31.78 nJ	34.19 nJ	34.49 nJ	28.19 nJ	28.50 nJ	30.77 nJ	30.71 nJ
4×4 SFSD, QPSK	24.33 nJ	24.24 nJ	26.07 nJ	26.29 nJ	21.75 nJ	21.98 nJ	23.74 nJ	23.68 nJ
4×4 SFSD, 16-QAM	53.76 nJ	53.54 nJ	57.61 nJ	58.11 nJ	48.21 nJ	48.68 nJ	52.69 nJ	52.52 nJ
LDPC $z = 32, R = 5/6$	6.838 μJ	6.814 μJ	7.332 μJ	7.394 μJ	5.415 μJ	5.475 μJ	5.703 μJ	5.690 μJ
LDPC $z = 32, R = 2/3$	6.923 μJ	6.895 μJ	7.423 μJ	7.486 μJ	5.608 μJ	5.667 μJ	5.924 μJ	5.916 μJ
LDPC $z = 64, R = 5/6$	13.288 μJ	13.243 μJ	14.249 μJ	14.369 μJ	12.176 μJ	12.310 μJ	13.289 μJ	13.257 μJ
LDPC $z = 64, R = 2/3$	13.440 μJ	13.383 μJ	14.411 μJ	14.534 μJ	12.309 μJ	12.440 μJ	13.433 μJ	13.415 μJ
LDPC $z = 96, R = 5/6$	20.456 μJ	20.386 μJ	21.935 μJ	22.120 μJ	18.238 μJ	18.439 μJ	19.906 μJ	19.857 μJ
LDPC $z = 96, R = 2/3$	21.207 μJ	21.117 μJ	22.740 μJ	22.933 μJ	19.528 μJ	19.735 μJ	21.311 μJ	21.282 μJ

Table 7.6: Total energy consumption for the implemented algorithms on 512-bit and 1024-bit SIMD processors.

Algorithm	512-bit SIMD				1024-bit SIMD			
	Bfy1	Cross1	Bfy2	Cross2	Bfy1	Cross1	Bfy2	Cross2
32-pt. FFT	7.26 nJ	6.58 nJ	6.61 nJ	6.23 nJ				
64-pt. FFT	13.58 nJ	13.63 nJ	14.00 nJ	13.75 nJ	17.55 nJ	15.78 nJ	15.77 nJ	15.21 nJ
128-pt. FFT	31.02 nJ	31.14 nJ	30.91 nJ	30.47 nJ	35.26 nJ	32.22 nJ	32.80 nJ	32.51 nJ
256-pt. FFT	64.96 nJ	65.21 nJ	70.19 nJ	70.07 nJ	72.13 nJ	72.90 nJ	72.78 nJ	72.48 nJ
512-pt. FFT	142.78 nJ	143.34 nJ	154.92 nJ	154.65 nJ	169.23 nJ	171.03 nJ	160.61 nJ	160.69 nJ
1024-pt. FFT	343.65 nJ	344.99 nJ	374.83 nJ	374.17 nJ	338.94 nJ	342.54 nJ	351.34 nJ	353.97 nJ
2048-pt. FFT	685.09 nJ	687.76 nJ	747.37 nJ	746.06 nJ	733.10 nJ	740.89 nJ	764.35 nJ	770.06 nJ
384-pt. FFT	113.50 nJ	113.94 nJ	121.19 nJ	120.97 nJ	137.18 nJ	138.64 nJ	145.20 nJ	146.28 nJ
768-pt. FFT	247.44 nJ	248.40 nJ	265.36 nJ	264.89 nJ				
4×4 FSD, QPSK	15.70 nJ	15.76 nJ	17.52 nJ	17.58 nJ	16.02 nJ	16.19 nJ	17.83 nJ	18.03 nJ
4×4 FSD, 16-QAM	26.83 nJ	26.94 nJ	29.94 nJ	30.04 nJ	27.37 nJ	27.66 nJ	30.47 nJ	30.81 nJ
4×4 SFSD, QPSK	20.59 nJ	20.67 nJ	22.96 nJ	23.01 nJ	21.02 nJ	21.17 nJ	23.42 nJ	23.76 nJ
4×4 SFSD, 16-QAM	45.42 nJ	45.57 nJ	50.64 nJ	50.81 nJ	46.25 nJ	46.40 nJ	51.28 nJ	52.05 nJ
LDPC $z=32, R=5/6$	4.950 μJ	4.969 μJ	5.522 μJ	5.541 μJ	5.088 μJ	5.088 μJ	5.692 μJ	5.720 μJ
LDPC $z=32, R=2/3$	5.173 μJ	5.193 μJ	5.771 μJ	5.786 μJ	5.300 μJ	5.320 μJ	5.932 μJ	5.982 μJ
LDPC $z=64, R=5/6$	10.151 μJ	10.228 μJ	10.932 μJ	10.966 μJ	10.029 μJ	10.123 μJ	11.179 μJ	11.335 μJ
LDPC $z=64, R=2/3$	10.479 μJ	10.398 μJ	11.308 μJ	11.339 μJ	10.456 μJ	10.580 μJ	11.626 μJ	11.824 μJ
LDPC $z=96, R=5/6$	17.971 μJ	18.038 μJ	20.063 μJ	20.126 μJ	18.902 μJ	18.752 μJ	21.065 μJ	20.789 μJ
LDPC $z=96, R=2/3$	18.767 μJ	18.841 μJ	20.931 μJ	20.990 μJ	19.633 μJ	19.299 μJ	21.816 μJ	21.650 μJ

Chapter 7 Evaluation of the SIMD architecture efficiency

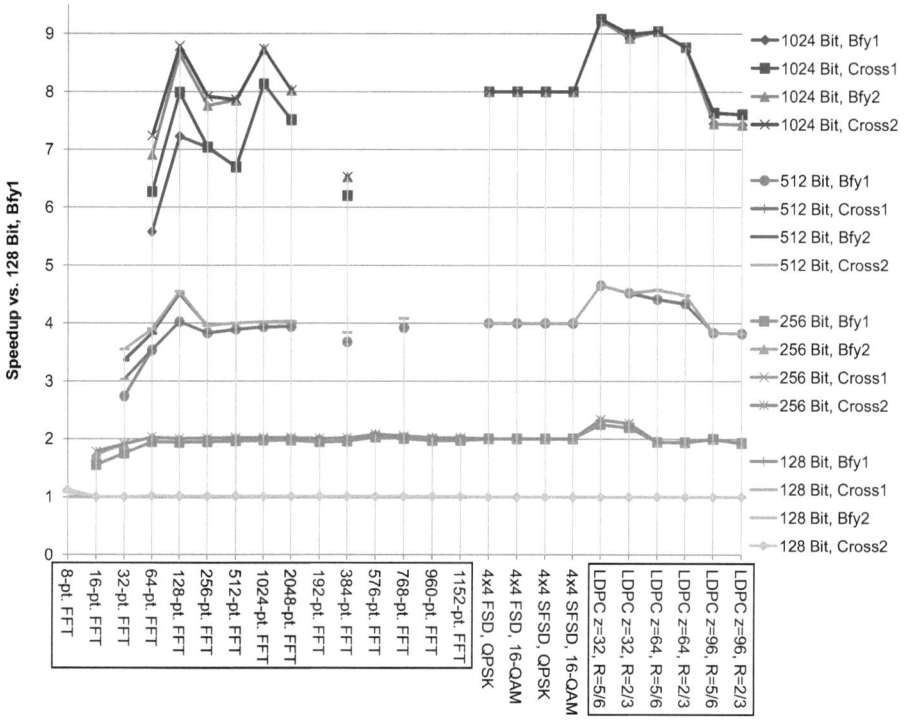

Figure 7.3: Speedup of algorithms compared to 128-bit SIMD processor with single-vector inverse butterfly network

The differences in the power consumption of crossbar and inverse butterfly networks for the same network width and the same SIMD width are minimal for most algorithms.

7.2.2 Energy-delay product analysis

The normalized energy-delay product is another metric that can be used to assess the efficiency of a processor architecture [GH96]. The normalized energy-delay product is a product of the normalized runtime and the normalized energy consumption. Normalization is again done by the results for a 128-bit SIMD processor with a single-vector inverse butterfly network.

The diagrams in figure 7.5, figure 7.6, and figure 7.7 show normalized energy-delay product *and* normalized area. The normalized energy-delay product is shown on the left ordinate and the normalized area on the right ordinate. SIMD widths are shown on the abscissa. The various curves represent different permutation networks.

Figure 7.5 shows normalized energy-delay curves for FFTs. The energy-delay product has been calculated by averaging energy and runtime for all FFT sizes. Figure 7.6 depicts normalized results

Chapter 7 Evaluation of the SIMD architecture efficiency

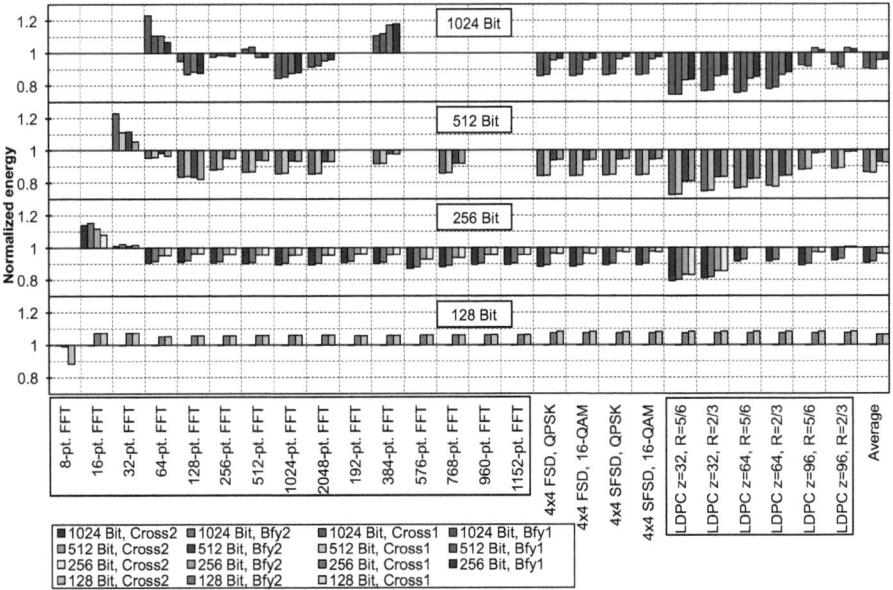

Figure 7.4: Normalized energy consumption of the implemented algorithms

for fixed-complexity sphere decoding, which also have been obtained by averaging. The normalized average energy-delay product for LDPC decoding is displayed in figure 7.7.

The normalized energy-delay product curves in figure 7.5 show that more complex permutation networks achieve better energy-delay products for the FFT processing on wider SIMD processors. Single-vector networks outperform double-vector networks for a 128-bit SIMD width, yet for wider SIMD widths, the double-vector networks achieve better energy-delay products. The smallest energy-delay product is achieved for a 1024-bit SIMD processor with a double-vector crossbar network. Yet, this processor also has the largest chip area.

The FSD and LDPC decoding results in figures 7.6 and 7.7 show a different behavior: As there is no performance gain for double-vector permutation networks compared to single-vector permutation networks, the energy-delay products for double-vector networks are always worse than the energy-delay products for single-vector networks.

Based on the analysis of all three algorithm classes, a single-vector network probably achieves the best efficiency in a wide SIMD LIW architecture for physical layer processing. The differences in the energy-delay products of crossbar and inverse butterfly networks are negligible, yet crossbar networks require more area, but also offer more flexibility. Hence, a decision between single-vector crossbar network and single-vector inverse butterfly network should be taken based on the requirements of algorithms.

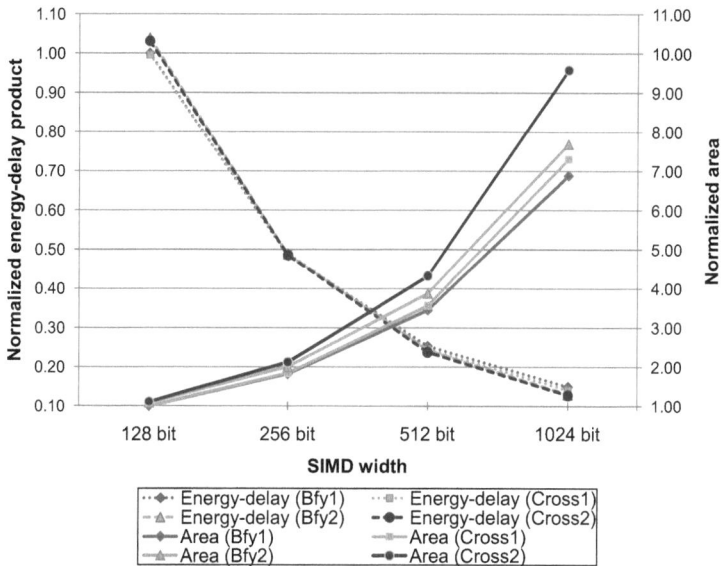

Figure 7.5: Normalized energy-delay product and area for radix-2 and mixed-radix FFTs

7.3 Possible approaches for improving the scalability

SIMD processing can be limited by available data parallelism, the overhead for aligning data on vectors, and the need for intra-vector calculations on segments of data vectors. Limited data parallelism cannot be compensated by changes on the processor architecture, but the other two issues can be addressed by modifications on the SIMD data path. These modifications do not provide unlimited scalability, yet the range of SIMD widths that still allows close to linear speedups can be extended and/or the algorithm complexity can be reduced.

The following two sections discuss indirect SIM_dD processing, which can improve the vector alignment for short mixed-radix FFTs and LDPC decoding, and support of operations on vector segments, which is useful for the parallel processing of multiple $z \times z$ sub-matrices in LDPC decoding.

7.3.1 Vector alignment with indirect SIM_dD processing

Indirect SIM_dD processing, as discussed in chapter 2.3.5, extends a normal SIMD data path with independent memory access capabilities for fixed vector segments with V_S elements ($V = k \cdot V_S$). Independent memory access is enabled by supporting multiple parallel address generation units and splitting one wide memory port into multiple smaller ports.

Indirect SIM_dD support is useful for LDPC decoding, when multiple sub-matrices in a row of the block matrix $\mathbf{H}_b$ are processed in parallel. One issue that limits the effectiveness of the parallel processing of multiple sub-matrices is the presence of empty sub-matrices (see chapter 6.3.3). If the expansion

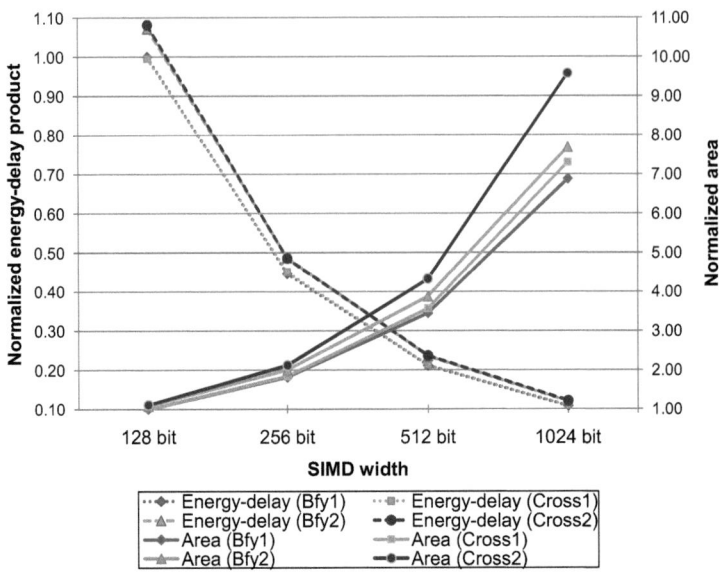

Figure 7.6: Normalized energy-delay product and area for soft-decision and hard-decision FSD

factor z of the quasi-cyclic LDPC code is a multiple of or equal to the segment size V_S in an indirect SIM$_d$D architecture, memory access for empty sub-matrices can be avoided. Furthermore, if sub-matrices from independent code words are processed in parallel, the required block interleaving of data in vectors can be done by memory access instead of requiring an additional permutation stage. As discussed in section 4.5.4, short mixed-radix FFTs can be parallelized by virtually reducing the vector length, i.e. data values that do not have to be processed together in the next FFT stages or data values from different FFTs are block-interleaved in vectors. The required permutations have a low complexity, yet many registers are required for storing independent data vectors, which reduces the number of consecutive FFT stages that can be processed without memory access. On an indirect SIM$_d$D processor architecture, the block interleaving can be done during the memory access; hence, the processing of FFT stages is no longer limited by available registers and the performance does not degrade.

The constraints for efficient SIMD processing of mixed-radix FFTs can be relaxed: If independent FFTs can be processed in parallel, the constraint that the FFT size is a multiple of the squared segment size ($N_{DFT} = k \cdot V_S^2$) is sufficient. If the virtual reduction of the vector length is done by block interleaving independent parts of one FFT, the additional constraint that the FFT size is a multiple of twice the vector length (as in the radix-2 FFT case) also must be fulfilled. If the whole mixed-radix FFT fits into the available SIMD registers, the block interleaving still has to be performed by permutation operations on the VPU.

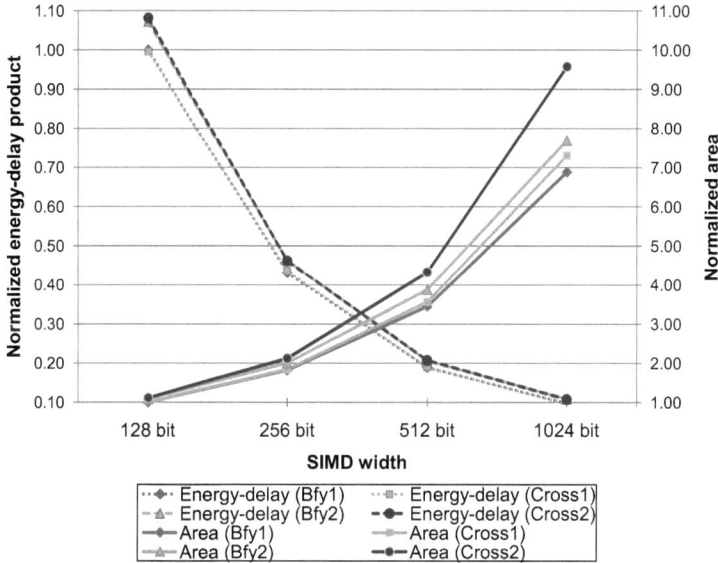

Figure 7.7: Normalized energy-delay product and area for the decoding of quasi-cyclic LDPC codes

7.3.2 Support for operations on vector segments

The parallel processing of multiple sub-matrices during the LDPC decoding is limited by the necessary reordering of vector elements. As cyclic shifts on sub-matrices have to be performed, independent permutations on vector segments have to be realized. On an SIMD processor with an inverse butterfly permutation network, the required permutations have to be defined by permutation patterns that are stored in memory. If the expansion factor z is greater than — yet not a multiple of — the vector length, data from multiple consecutive vectors also has to be merged on a vector segment basis, which requires complex vector masks that also have to be stored in memory. If the required operations on vector segments are directly supported by instructions, the memory requirements can be reduced and the performance improves.

If sub-matrices from independent code words are processed in parallel in vector segments, the same operations (permutations, merging by masked move operations) have to be done for all vector segments. The required changes are supporting rotations on vector segments, with the same rotation distance for all segments, and supporting an instruction, which initializes all segments in a vector mask by setting the first k bits to one. The latter can be implemented with some additional control logic compared to an instruction, which sets the first k bits of the complete vector to one. A rotation on vector segments can also be easily implemented on an inverse butterfly network. On an inverse butterfly network, rotations are computed recursively by rotations on parts of a vector. For example, the first $\log_2(V) - 1$ perform the same rotation operation on the two halves of a vector. The last butterfly stage permutes elements from both vector halves to compute the rotation on V element.

Hence, rotations on vector segments can be computed by disabling the last butterfly stage or stages, e. g. by simply setting all control signals to zero. An example for a vector length of 16 elements and a cyclic shift of three elements is displayed in figure 7.8.

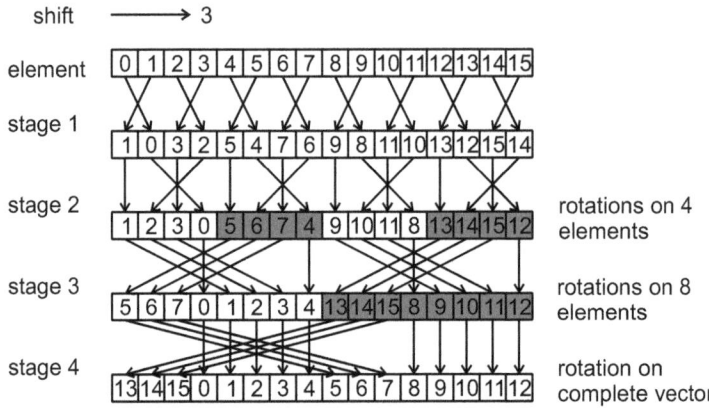

Figure 7.8: Cyclic shift operation on an inverse butterfly network. The complete cyclic shift is computed from cyclic shifts on smaller vector segments.

If sub-matrices from one code word are processed in parallel in vector segments, the cyclic shift distances are no longer the same for all vector segments. Therefore, rotation operations that take multiple scalar rotation distances as an input, one for each vector segment, are required, which increases the hardware complexity. Hence, a parallel processing of independent code words is the preferred solution for LDPC decoding on wide SIMD processors for SIMD widths beyond the expansion factor of the quasi-cyclic code.

7.4 Software development for LIW SIMD processors

One important issue for the development of SDR systems based on SIMD processors is mapping applications on the processor architecture. In the best case, the software should be written in a high-level language (e. g. C) and mapped on the processor by an optimizing compiler. According to a commonly used rule of thumb, compiler assisted software development should ideally lead to 80 percent of the performance of hand-optimized assembly code with 20 percent of the effort.

In the context of SIMD processors for SDR with LIW support, the task of the compiler can be characterized by three subtasks:

1. Operations in a high-level language have to be mapped on DSP instructions. Dependent on the instruction set and data types, this task includes the mapping on complex multi-stage instructions, for example an add-compare-select instruction for Viterbi decoding, and the detection of rounding and saturation logic.

2. Vectorizable program segments have to be recognized and efficiently mapped on SIMD instructions, the difficulties of this crucial optimization step are discussed in the following.

3. Operands have to be assigned to registers and instructions have to be scheduled in LIW instructions. The efficiency of this task depends largely on the partitioning of the algorithm into loops and functions.

All three subtasks still require programmers with detailed knowledge about the processor's instruction set architecture (ISA). Furthermore, the input programs have to be written in a manner that the compiler can understand.

For a SIMD processor architecture, the vectorization of algorithms is the most important — and most challenging — issue. Compiler-based vectorization, e.g. for GPPs with short SIMD extensions [NZ06, NRZ06] is usually done by analyzing inner program loops. Data dependencies of array variables, which might prevent vectorization, are analyzed and predefined loop structures, e.g. reduction operations that sum up array elements, are recognized. If an inner loop is vectorizable, the code is replaced by vector operations. Due to the complexity of data access patterns in DSP algorithms, this approach is insufficient for wide SIMD SDR processors. For example, vectorizing an FFT requires first choosing a vectorizable algorithm (see chapter 4). On the other hand, the FSD algorithm allows different vectorization strategies, which have different overheads for vector permutation operations. Yet, the optimal strategy cannot be detected by analyzing inner loops separately.

Regular transforms, like the FFT, can be vectorized using a representation by Kronecker products [Pit97]. For example, the SPIRAL compiler [PMJ$^+$05, FP02] uses a mathematical description of algorithms as its input. The algorithm is vectorized using formula manipulations and the design space of possible algorithm decompositions is automatically explored. Yet, this approach is not applicable to arbitrary algorithms.

One possible approach for vectorizing algorithms that cannot be expressed by regular transforms is extending the traditional approach for the vectorization of inner loops with an analysis of the vector data alignment for the complete program [WSK07, WSK08]: As a first step, all options for parallelizing nested loops are explored. Next, the costs for aligning data (i.e. the overhead for vector permutations) in loops and realigning data between loop nests have to be estimated. The cost estimation is a complex task for LIW SIMD processors, as the performance is only degraded by permutation operations if the permutations cannot be done in parallel to useful arithmetical operations in LIW instructions. Yet, the mapping of loop code on LIW instructions in turn depends on the vectorization. If an accurate cost model for data alignment is available, the mapping on vector operations can be done by solving an optimization problem on a graph that describes the different possible vectorization strategies and their costs.

As the complexity of the design of an optimizing compiler for SIMD SDR processors is very high, most wide SIMD processor architectures only support some subtasks of the software development by automatic tools, the remaining tasks have to be done by the programmers: The EVP is programmed in EVP-C [vHM$^+$04]; a C-based language with specialized commands, called intrinsics, for SIMD vector operations and DSP instructions, the compiler only performs register allocation and optimization for VLIW processing. The Ardbeg processor's instruction set is based on the ARM NEON SIMD instruc-

Chapter 7 Evaluation of the SIMD architecture efficiency

Table 7.7: Comparison of compiler generated code on the EVP and hand-optimized assembly code on the proposed SIMD architecture (256 bit SIMD width, crossbar network). FFT throughput is measured in FFTs per second, while FSD throughput is measured in OFDM subcarriers per second.

Algorithm	Throughput on SIMD core	Throughput on EVP	Difference [%]
64-pt. FFT	$6.00 \cdot 10^6$ FFTs/s	$6.00 \cdot 10^6$ FFTs/s	$\pm 0\%$
256-pt. FFT	$1.14 \cdot 10^6$ FFTs/s	$1.07 \cdot 10^6$ FFTs/s	-6.1 %
512-pt. FFT	$5.07 \cdot 10^5$ FFTs/s	$4.81 \cdot 10^5$ FFTs/s	-5.1 %
1024-pt. FFT	$2.08 \cdot 10^5$ FFTs/s	$2.11 \cdot 10^5$ FFTs/s	+1.4 %
2048-pt. FFT	$1.04 \cdot 10^5$ FFTs/s	$9.77 \cdot 10^4$ FFTs/s	-6.1 %
192-pt. FFT	$1.46 \cdot 10^6$ FFTs/s	$1.39 \cdot 10^6$ FFTs/s	-4.8 %
384-pt. FFT	$6.76 \cdot 10^5$ FFTs/s	$6.15 \cdot 10^5$ FFTs/s	-9.0 %
576-pt. FFT	$3.82 \cdot 10^5$ FFTs/s	$3.55 \cdot 10^5$ FFTs/s	-7.1 %
768-pt. FFT	$2.95 \cdot 10^5$ FFTs/s	$2.71 \cdot 10^5$ FFTs/s	-8.1 %
960-pt. FFT	$2.11 \cdot 10^5$ FFTs/s	$1.85 \cdot 10^5$ FFTs/s	-12.3 %
1152-pt. FFT	$1.80 \cdot 10^5$ FFTs/s	$1.69 \cdot 10^5$ FFTs/s	-6.1 %
FSD channel ordering	$1.24 \cdot 10^7$ FFTs/s	$1.18 \cdot 10^7$ FFTs/s	-4.8 %
QR-decomposition	$1.14 \cdot 10^7$ FFTs/s	$9.60 \cdot 10^6$ FFTs/s	-15.8 %
QPSK FSD search	$1.89 \cdot 10^7$ FFTs/s	$1.62 \cdot 10^7$ FFTs/s	-14.3 %
16-QAM FSD search	$4.77 \cdot 10^6$ FFTs/s	$3.75 \cdot 10^6$ FFTs/s	-21.4 %

tion set [WLS+08a]. The NEON instructions are supported by intrinsics in the compiler [ARM08]. The Sandblaster processor is the only SIMD SDR processor with a complete vectorizing compiler, which also supports multi-threading [JGKM04]. However, as reported in [JSJ09], the vectorization does not always work.

The programming model that has been selected for the EVP offers performance that is close to the performance of hand-coded assembly code. Yet as the vectorization, the mapping on intrinsics, and the decomposition of the algorithm into loops that can be efficiently processed by the VLIW compiler still have to be done manually, the required programming effort and knowledge of the processor architecture are still high. The performance of the EVP-C compiler has been evaluated by comparing the results to the performance of hand-coded assembly code on the proposed SIMD processor architecture for FFT and sphere decoding algorithms. Performance results for the same vector length and the same permutation network topology have been compared. As the EVP supports more LIW slots, the comparison is not 100 percent accurate. The results are listed in table 7.7.

For most considered algorithms, the performance degradation due to the compiler is less than 10 percent. Only complex loops, e. g. the radix-5 DFT stage in the 960-point FFT and the FSD tree search, achieve significantly worse performance.

Chapter 8
Conclusion

4G SDR systems come with steep requirements on the efficiency of programmable architectures. SIMD processors can potentially achieve a good ratio between energy consumption and throughput performance. Yet, SIMD processors require algorithms that can utilize all parallel data lanes efficiently. Therefore, the scalability of SIMD processing for key algorithms of 4G wireless systems has been investigated in this book.

Overview of results

A scalable SIMD processor architecture has been proposed in chapter 3, which enables an exploration of SIMD widths ranging from 128 bit to 1024 bit as well as four different permutation networks for vector element permutations. The considered permutation networks are single-vector inverse butterfly, single-vector crossbar, double-vector inverse butterfly, and double-vector crossbar networks. Supplying different permutation networks allows investigating how the support of more complex permutations influences the algorithm performance and the complexity of the processor architecture. LIW execution has also been implemented, as the parallel processing of computational operations (e. g. a vector addition) and overhead operations (e. g. vector element permutations) improves the scalability by hiding overhead operations. Synthesized gate level processor models for all SIMD processors allow estimating chip area, maximum frequency, power consumption, and energy consumption.

Radix-2 and mixed-radix FFT algorithms for SIMD processors have been investigated in chapter 4. FFT algorithms that enable efficient SIMD processing have been derived from the matrix representation of the FFT. The radix-2 FFT algorithm requires that the FFT size is at least twice the SIMD vector length V, yet the mixed-radix FFT algorithm only enables efficient vector processing for FFT sizes that are a multiple of the squared vector length. Both algorithms perform all FFT stages (e. g. radix-2 butterfly stages) on complete data vectors and require only $\log_2(V)$ permutation stages that can mostly be realized by simple masked butterfly permutations on pairs of vectors. If the constraints are satisfied, close to linear speedup can be achieved. The achieved throughput performance is competitive to the performance of dedicated FFT processors.

Sphere decoding, which is a technique for ML detection in spatial multiplexing MIMO systems, has been studied in chapter 5. The original sphere decoder (SD) algorithm is a sequential algorithm with a variable complexity, i. e. the number of required iterations depends on the input data. Hence, the SD algorithm is neither suited for real-time hardware implementations nor for SIMD processing.

Chapter 8 Conclusion

Therefore, sphere decoding algorithms that combine a fixed-complexity with a parallelizable structure have emerged. The fixed-complexity sphere decoder (FSD) has been mapped on the scalable SIMD processor architecture, as the algorithm has a relatively low computational complexity and achieves a BER performance close to the ML performance [BT06a, BT06c, BT08b]. The FSD algorithm also has been extended to a soft-decision algorithm, which computes log-likelihood ratios. The algorithm performance scales linearly with the SIMD width if multiple sub-carriers of an OFDM symbol are processed in parallel. The achieved throughput for soft-decision output on a 1024-bit SIMD processor is greater than 100 Mbps for a 4×4 MIMO-OFDM system with 16-QAM modulation; hence, the throughput should be sufficient for most battery-powered mobile devices in future 4G systems. The throughput performance is better than the performance of other reported SDR sphere decoding implementations, but SIMD processors achieve neither the high throughput nor the high energy efficiency of a state-of-the-art MIMO detection ASIC.

The decoding of quasi-cyclic LDPC codes has been investigated in chapter 6. Quasi-cyclic codes are a class of structured LDPC codes, which are described by $z \times z$ sub-matrices that are either empty or cyclic shifted unity matrices. The decoding of LDPC codes is done by message-passing algorithms on a graph-representation of the parity check matrix, which consists of bit nodes and (parity) check nodes. The regular structure of quasi-cyclic codes enables processing z bit nodes and z check nodes in parallel. Implementation results show that linear or close to linear speedups can be achieved on wider SIMD processors even for SIMD widths greater than the expansion factor. Yet, in this case, the processing is more elaborate, as cyclic shifts on segments of data vectors are necessary. The maximum throughput that has been achieved is 36.59 Mbps for a rate 5/6 WiMAX code with $z = 64$ on a 1024-bit SIMD processor. The throughput could be easily improved by supporting 8-bit data types next to 16-bit data types for vector elements and by adding custom instructions for the check node processing. In the current state, both throughput performance and energy consumption are significantly worse than throughput performance and energy consumption of ASIC solutions for LDPC decoding.

The energy efficiency and the chip area of the proposed scalable SIMD processor architecture have been analyzed in chapter 7. Results for energy consumption and energy-delay product show that the energy efficiency increases with the SIMD width, as the relative amount of energy spent on the decoding of instructions and the scalar data path decreases. Single-vector permutation networks achieve the best energy-delay product results for sphere decoding and LDPC decoding. Double-vector permutation networks achieve slightly better energy-delay products than single-vector permutation networks for FFTs on 512-bit and 1024-bit SIMD processors. Overall, single-vector permutation networks appear to be the best option for SIMD processors with LIW support for SDR algorithms. Networks with inverse butterfly and crossbar topologies achieve similar energy efficiency results. Crossbar networks offer more flexibility at the cost of more chip area.

Selection of SIMD width and permutation network topology

The optimal combination of SIMD width and permutation network topology depends on the targeted application(s). Yet, some basic guidelines can be derived from the design space exploration in this

Chapter 8 Conclusion

paper. The main criterion for selecting the SIMD width is the desired throughput performance of an application. The achievable throughput depends on the computational complexity of algorithms on the SIMD architecture. The complexity of FFT algorithms is the lowest of all three considered algorithm classes. Yet, radix-2 and especially mixed-radix FFTs also require the most complex data alignment. The most complex task is LDPC decoding, which is more than one order of magnitude more complex than FFT processing. The complexity of sphere decoding using the FSD algorithm is somewhere in the middle between the other two algorithm classes.

Due to the high complexity of LDPC decoding, a SIMD-based architecture for LDPC decoding and MIMO-OFDM should be a multi-core processor architecture. Such an architecture would comprise at least one SIMD processor dedicated to MIMO-OFDM processing, i.e. sphere decoding and FFT processing, as well as other less demanding tasks, and several SIMD processors optimized for LDPC decoding.

The SIMD width for LDPC decoding should be as wide as possible, as wide SIMD processors in principle achieve the best energy efficiency figures. If quasi-cyclic LDPC codes with expansion factors that are not multiples of the SIMD width shall be decoded, a crossbar network topology offers better performance than an inverse butterfly network topology. Alternatively, a modified inverse butterfly network with support for permutations on vector segments could be used (see section 7.3.2). Compared to single-vector permutation networks, double-vector permutation networks offer no performance gain, but they require more energy for the register files. Therefore, the permutation network should be a single-vector network.

For MIMO-OFDM processing, the optimal SIMD architecture depends on the concrete algorithm mix and the desired throughput. In principle, the alternatives are using many SIMD cores with short SIMD widths or few wide SIMD cores. Wide SIMD processors potentially offer better energy efficiency, but also are less flexible concerning the alignment of data values, as longer data vectors are processed. In the following, an example for a hypothetical cellular MIMO-OFDM transmission system using OFDM-A for the downlink (DL) and SC-FDMA for the uplink (UL) channel is used to illustrate the decision process for the design of a SIMD-based architecture for a mobile device. The scenario assumes 4×4 MIMO with 16-QAM modulation, an OFDM symbol size of 1024 sub-carriers, and 10,000 OFDM symbols per second. In the DL channel (OFDM-A), at most 600 sub-carriers are assigned to one user, while at most 384 sub-carriers are assigned to one user in the UL channel (SC-FDMA). This corresponds to peak throughputs of 96 Mbps in the DL and 61.44 Mbps in the UL channel. The main tasks that have to be performed on the SIMD-based architecture are radix-2 and mixed-radix FFTs for UL SC-FDMA modulation, radix-2 FFTs for DL OFDM-A demodulation and sphere decoding using the FSD algorithm in the DL channel.

Figures 8.1 and 8.2 list processor utilization[1] and total energy consumption results, respectively, for different multi-core SIMD architectures that can meet the required throughput performance. The energy consumption figures include only the processor cores, but neither interconnect nor memories.

---

[1] The processor utilization describes the relative amount of time spend on a task on the multi-core architecture. At a processor utilization of 100 percent, all processor cores are occupied all the time and no further tasks can be executed without adding further processor cores.

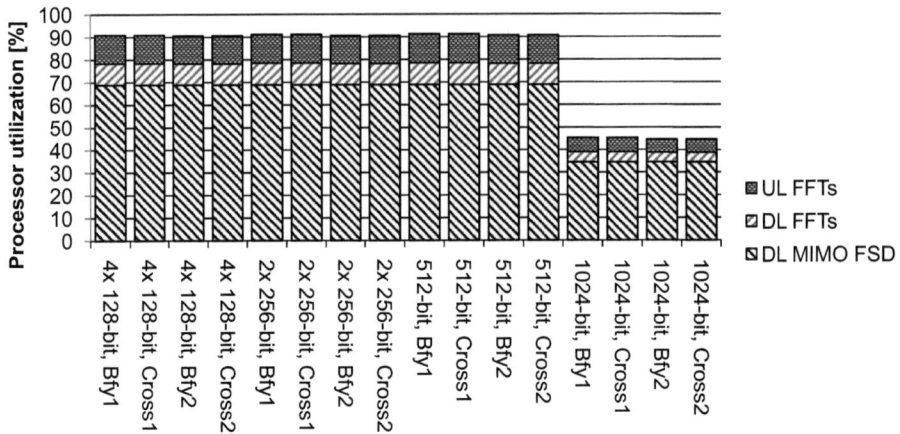

Figure 8.1: Processor utilization for the example MIMO-OFDM transmission scenario

As figure 8.1 shows, a single 512-bit SIMD processor can achieve the required throughput performance at a total processor utilization of about 90 percent, the same performance can be reached with two 256-bit SIMD processors or four 128-bit SIMD processors. One 1024-bit SIMD processor can also achieve the required throughput performance at approximately 45 percent processor utilization. However, a 1024-bit SIMD processor is oversized for this scenario, as illustrated by the area curve in figure 8.3.

Figure 8.2 shows that the 512-bit SIMD processor architectures are the most energy efficient processors for this scenario.[2] Single-vector permutation networks require significantly less energy and achieve almost the same performance as double-vector permutation networks. Therefore, a single-vector permutation network should be selected. Single-vector inverse butterfly and crossbar networks achieve similar processor utilization and energy consumption results, but inverse butterfly networks require less area (see figure 8.3). Hence, for this scenario, a 512-bit SIMD processor architecture with a single-vector inverse butterfly network is an adequate solution. If increased flexibility for vector data alignment is desired, two 256-bit SIMD processors could instead be used at the cost of increased energy consumption.

Possible topics for future research

The performance of short mixed-radix FFTs, whose sizes are not multiples of the squared vector length, and the LDPC decoding for vector lengths greater than the expansion factor z is limited due to the complexity of the required permutations and memory access patterns. As discussed in chapter 7.3, indirect SIM_dD processing and efficient support for parallel permutation operations on segments

---

[2]The 1024-bit SIMD results are slightly worse due to the RTL code automatically generated from LISA.

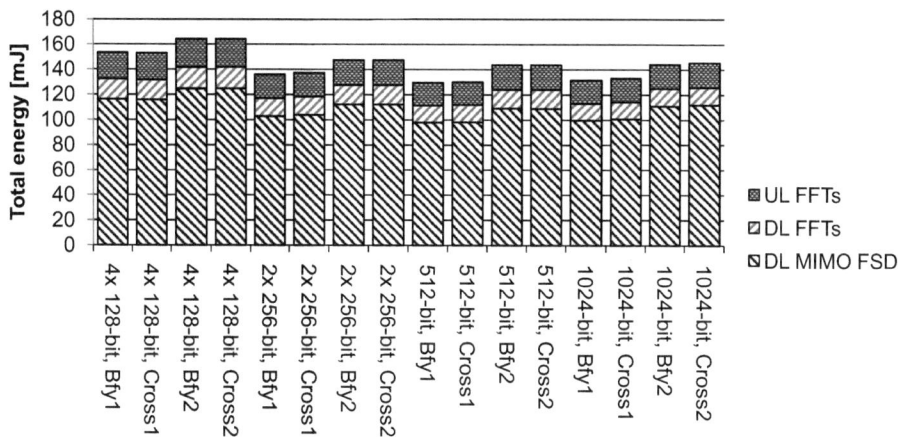

Figure 8.2: Total energy consumption for the example MIMO-OFDM transmission scenario

of vectors are two techniques that could improve the scalability and reduce the complexity of algorithm implementations. Hence, these techniques merit further investigation in future research.

Another open topic for research is the development of efficient vectorizing compilers for wide SIMD processors. Currently, the most appropriate approach is performing the vectorization and the mapping of operations on DSP instructions manually by programming in a high-level language with support of vector operation intrinsics (e. g. as in EVP-C, the programming language for the EVP). Yet, the availability of good vectorizing compilers would reduce the complexity of the software development and improve the acceptance of wide SIMD processors.

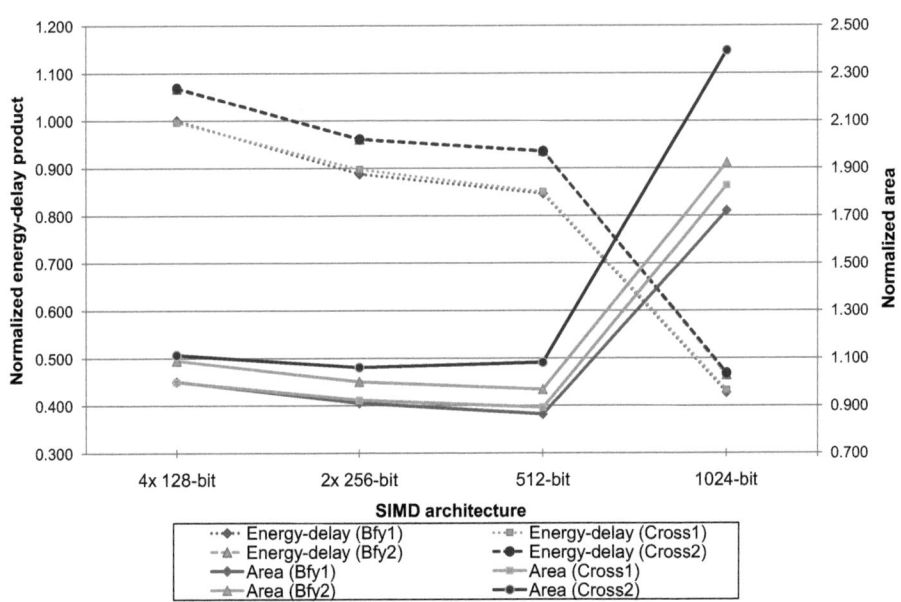

Figure 8.3: Normalized energy-delay product and area for the described scenario

Bibliography

[AG09]   Inc. Athena Group. FFT Cores - FFTs for Communications and Signal Processing. Online, http://www.athena-group.com/external/FFT_PB_Book.pdf, 2009.

[Ala98]   S. M. Alamouti. A Simple Transmit Diversity Technique for Wireless Communications. *IEEE J. Select. Areas Commun.*, 16(8):1451–1458, 1998.

[ARB+05] R. Azevedo, S. Rigo, M. Bartholomeu, G. Araujo, C. Araujo, and E. Barros. The ArchC architecture description language and tools. *Int. J. Parallel Program.*, 33(5):453–484, October 2005.

[ARC05]   ARC International. ARChitect Processor Configurator: The Power of Configurable Processing At Your Fingertips. Whitepaper, online http://www.arc.com/configurablecores/architect/, 2005.

[ARM08]   ARM. ARM NEON support in the ARM compiler. White paper, http://www.arm.com/files/pdf/NEON_Support_in_the_ARM_Compiler.pdf, Sept. 2008.

[Bö6]   Helmut Bölcskei. MIMO-OFDM wireless systems: Basics, perspectives, and challenges. *IEEE Wireless Communications*, 13:31–37, August 2006.

[Bau01]   G. Bauch. Introduction to multi-antenna systems and space-time codes. In *Proc. of the 3rd International Workshop on Commercial Radio Sensors and Communication Techniques*, 2001.

[BBW+05] A. Burg, M. Borgmann, M. Wenk, M. Zellweger, W. Fichtner, and H Bölcskei. VLSI implementation of MIMO detection using the sphere decoding algorithm. *IEEE J. Solid-State Circuits*, 40(7):1–12, July 2005.

[BD03]   L. Barnault and D. Declercq. Fast decoding algorithm for LDPC over GF(2q). In *Information Theory Workshop, 2003. Proceedings. 2003 IEEE*, pages 70 – 73, 31 2003.

[BDT08]   BDTi. An Independent Evaluation of the picoChip PC102 Software Development Tools and Programming Model. Technical report, Berkeley Design Technology, Inc., 2008.

[Beh09]   B. Beheshti. On performance of LTE UE DFT and FFT implementations in flexible software based baseband processors. In *Systems, Applications and Technology Conference, 2009. LISAT '09. IEEE Long Island*, pages 1 –4, 1-1 2009.

[Ben65]   V. E. Beneš. *Mathematical theory of connecting networks and telephone traffic*. Academic Press, 1965.

Bibliography

[BGT93]   C. Berrou, A. Glavieux, and P. Thitimajshima. Near shannon limit error-correcting coding and decoding: Turbo-codes. 1. In *Communications, 1993. ICC 93. Geneva. Technical Program, Conference Record, IEEE International Conference on*, volume 2, pages 1064–1070, 23-26 1993.

[Bhu09]   M. R. Bhujade. *Parallel Computing*. New Age Publications (Academic), 2nd edition, 2009.

[BNW98]   M. Beneš, S. M. Nowick, and A. Wolfe. A Fast Asynchronous Huffman Decoder for Compressed-Code Embedded Processors. In *ASYNC '98: Proceedings of the 4th International Symposium on Advanced Research in Asynchronous Circuits and Systems*, pages 43–56, Washington, DC, USA, 1998. IEEE Computer Society.

[BT06a]   L. G. Barbero and J. S. Thompson. A Fixed-Complexity MIMO Detector Based on the Complex Sphere Decoder. In *IEEE International Workshop on Signal Processing Advances in Wireless Communications (SPAWC '06)*, Cannes, France, Jul. 2006.

[BT06b]   L.G. Barbero and J.S. Thompson. Performance Analysis of a Fixed-Complexity Sphere Decoder in High-Dimensional MIMO Systems. In *Acoustics, Speech and Signal Processing, 2006. ICASSP 2006 Proceedings. 2006 IEEE International Conference on*, volume 4, 14-19 2006.

[BT06c]   L.G. Barbero and J.S. Thompson. Rapid Prototyping of a Fixed-Throughput Sphere Decoder for MIMO Systems. In *Communications, 2006. ICC '06. IEEE International Conference on*, volume 7, pages 3082 –3087, June 2006.

[BT08a]   L.G. Barbero and J.S. Thompson. Extending a Fixed-Complexity Sphere Decoder to Obtain Likelihood Information for Turbo-MIMO Systems. *Vehicular Technology, IEEE Transactions on*, 57(5):2804 –2814, Sept. 2008.

[BT08b]   L.G. Barbero and J.S. Thompson. Fixing the Complexity of the Sphere Decoder for MIMO Detection. *Wireless Communications, IEEE Transactions on*, 7(6):2131–2142, June 2008.

[CDE+05] J. Chen, A. Dholakia, E. Eleftheriou, M.P.C. Fossorier, and X.-Y. Hu. Reduced-Complexity Decoding of LDPC Codes. *Communications, IEEE Transactions on*, 53(8):1288–1299, Aug. 2005.

[CEV09]   CEVA, Inc. CEVA-XC Communications Processor. Online, http://www.ceva-dsp.com/products/system/pdf/ceva-xc_datasheet.pdf, 2009.

[CFRU01] S.-Y. Chung, Jr. Forney, G.D., T.J. Richardson, and R. Urbanke. On the design of low-density parity-check codes within 0.0045 dB of the Shannon limit. *Communications Letters, IEEE*, 5(2):58 –60, Feb. 2001.

[CoW09a] CoWare. *CoWare Processor Designer Product Family: LISA Language Reference Manual*. CoWare Inc., product version v2009.1.1 edition, April 2009.

[CoW09b] CoWare. *CoWare Processor Designer Product Family: Processor Design Guide*. CoWare Inc., product version v2009.1.1 edition, April 2009.

[CoW09c] CoWare. *CoWare Processor Designer Product Family: Processor Generator Reference Manual*. CoWare Inc., product version v2009.1.1 edition, April 2009.

[CT65] J. W. Cooley and J. W. Tukey. An algorithm for the machine calculation of complex Fourier series. *Math. Comput.*, 19:297–301, 1965.

[CTC+04] A. Chun, E. Tsui, I. Chen, H. Honary, and J. Lin. Application of the Intel reg; reconfigurable communications architecture to 802.11a, 3G and 4G standards. In *Emerging Technologies: Frontiers of Mobile and Wireless Communication, 2004. Proceedings of the IEEE 6th Circuits and Systems Symposium on*, volume 2, pages 659 – 662, 31 2004.

[Dal90] W. J. Dally. Performance Analysis of k-ary n-cube Interconnection Networks. *IEEE Trans. Comput.*, 39(6):775–785, June 1990.

[dB87] V. L. de Buen. Multistage Interconnection Networks In Multiprocessor Systems. A Simulation Study. *Quaderns d'Estadística i Investigació Operativa*, 11(3):73–86, 1987.

[DM98] M.C. Davey and D.J.C. MacKay. Low density parity check codes over GF(q). In *Information Theory Workshop, 1998*, pages 70 –71, 22-26 1998.

[DM03] J.H. Derby and J.H. Moreno. A high-performance embedded DSP core with novel SIMD features. In *Acoustics, Speech, and Signal Processing, 2003. Proceedings. (ICASSP '03). 2003 IEEE International Conference on*, volume 2, pages 301–304, 6-10 2003.

[DMW03] J. H. Derby, J. H. Moreno, and M. S. Ware. A low-power high-performance embedded DSP core with novel SIMD features. In *GSPx Conference 2003*, 2003.

[DOW96] S. Dutta, K. J. O'Connor, and A. Wolfe. High-performance crossbar interconnect for a VLIW video signalprocessor. In *Proceedings., Ninth Annual IEEE International ASIC Conference and Exhibit*, pages 45–49, Sep. 1996.

[DWWO96] S. Dutta, A. Wolfe, W. Wolf, and K.J. O'Connor. Design issues for very-long-instruction-word VLSI video signal processors. In *VLSI Signal Processing, IX, 1996., [Workshop on]*, pages 95–104, 1996.

[FG98] G. J. Foschini and M. J. Gans. On limits of wireless communications in a fading environment when using multiple antennas. *Wireless Personal Communications*, 6:311–335, 1998.

[Fis83] J. A. Fisher. Very Long Instruction Word architectures and the ELI-512. In *ISCA '83: Proceedings of the 10th annual international symposium on Computer architecture*, pages 140–150, New York, NY, USA, 1983. ACM.

Bibliography

[FLN+09] R. Fasthuber, M. Li, D. Novo, P. Raghavan, L. Van Der Perre, and F. Catthoor. Novel energy-efficient scalable soft-output SSFE MIMO detector architectures. In *Systems, Architectures, Modeling, and Simulation, 2009. SAMOS '09. International Symposium on*, pages 165 –171, 20-23 2009.

[Fly66] M. J. Flynn. Very high-speed computing systems. *Proceedings of the IEEE*, 54(12):1901–1909, 1966.

[Fly72] M. J. Flynn. Some Computer Organizations and Their Effectiveness. *Computers, IEEE Transactions on*, C-21(9):948 –960, Sept. 1972.

[For07] SDR Forum. SDRF Cognitive Radio Definitions, Nov. 2007.

[Fos96] G. J. Foschini. Layered space-time architecture for wireless communication in a fading environment when using multi-element antennas. *Bell Labs Technical Journal*, pages 41–59, 1996.

[Fos04] M.P.C. Fossorier. Quasicyclic low-density parity-check codes from circulant permutation matrices. *Information Theory, IEEE Transactions on*, 50(8):1788 – 1793, Aug. 2004.

[FP85] U. Fincke and M. Pohst. Improved methods for calculating vectors of short length in a lattice, including a complexity analysis. *Mathematics of Computation*, 44(170):463–471, Apr. 1985.

[FP02] F. Franchetti and M. Püschel. A SIMD vectorizing compiler for digital signal processing algorithms. In *IPDPS '02: Proceedings of the 16th International Parallel and Distributed Processing Symposium*, 2002.

[FP03] F. Franchetti and M. Püschel. Short Vector Code Generation for the Discrete Fourier Transform. In *International Parallel and Distributed Processing Symposium (IPDPS'03)*, 2003.

[FP07] F. Franchetti and M. Püschel. SIMD vectorization of non-two-power sized FFTs. In *Proceedings of International Conference on Acoustics, Speech, and Signal Processing (ICASSP) 2007*, 2007.

[Fra03] F. Franchetti. *Performance Portable Short Vector Transforms*. PhD thesis, Vienna University of Technology, 2003.

[FVP07] F. Franchetti, Y. Voronenko, and M. Püschel. A rewriting system for the vectorization of signal transforms. *High Performance Computing for Computational Science - VECPAR 2006*, 4395/2007:363–377, 2007.

[Gal63] R. G. Gallager. *Low Density Parity Check Codes*. M.I.T. Press, 1963.

[GCT92] J. Granata, M. Conner, and R. Tolimieri. Recursive Fast Algorithms and the Role of the Tensor Product. *IEEE Trans. Signal Processing*, 40(12):2921–2930, Dec. 1992.

Bibliography

[Gen73] W. M. Gentleman. Least Squares Computations by Givens Transformations Without Square Roots. *IMA Journal of Applied Mathematics*, 12(3):329–336, 1973.

[GH96] R. Gonzalez and M. Horowitz. Energy dissipation in general purpose microprocessors. *Solid-State Circuits, IEEE Journal of*, 31(9):1277–1284, Sept. 1996.

[GI06] J. Glossner and D. Iancu. The Sandbridge SB3011 SDR platform. In *Proceedings of the Symposium on Trends in Communications (SympoTIC'06)*, Bratislava, Slovakia, 2006.

[GL73] L. R. Goke and G. J. Lipovski. Banyan networks for partitioning multiprocessor systems. *SIGARCH Comput. Archit. News*, 2(4):21–28, 1973.

[GLGVP06] G. Goossens, D. Lanneer, W. Geurts, and J. Van Praet. Design of ASIPs in multiprocessor SoCs using the Chess/Checkers retargetable tool suite. In *System-on-Chip, 2006. International Symposium on*, pages 1–4, Nov. 2006.

[GM05] Y. Guo and D. McCain. Reduced QRD-M detector in MIMO-OFDM systems with partial and embedded sorting. In *Global Telecommunications Conference (GLOBECOM '05)*, 2005.

[GN06] Z. Guo and P. Nilsson. Algorithm and implementation of the K-best sphere decoding for MIMO detection. *Selected Areas in Communications, IEEE Journal on*, 24(3):491–503, Mar. 2006.

[Goo58] T. J. Good. The interaction algorithm and practical Fourier analysis. *Journal of the Royal Statistical Society, Series B (Methodological)*, 20(2):361–372, 1958.

[GRS07] C. Grassmann, M. Richter, and M. Sauermann. Mapping the physical layer of radio standards to multiprocessor architectures. In *Design, Automation Test in Europe Conference Exhibition, 2007. DATE '07*, pages 1–6, April 2007.

[GS91] J. Götze and U. Schwiegelshohn. A square root and division free Givens rotation for solving least squares problems on systolic arrays. *SIAM J. Sci. Stat. Comput.*, 12(4):800–807, 1991.

[GVL96] Gene H. Golub and Charles F. Van Loan. *Matrix computations (3rd ed.)*. Johns Hopkins University Press, Baltimore, MD, USA, 1996.

[GZMC06] Y. Guo, J. Zhang, D. McCain, and J. R. Cavallaro. An efficient circulant MIMO equalizer for CDMA downlink: Algorithm and VLSI architecture. *EURASIP J. Appl. Signal Process.*, 2006:1–18, 2006.

[GZYC86] Q. Gao, X. Zhang, S. Yang, and S. Chen. Vector computer 757. *Journal of Computer Science and Technology*, 1(3):1–14, Sept. 1986.

[Hie03] T. Hiers. TMS320C6414/5/6 Power Consumption Summary - Application Report. Technical report, Texas instruments, 2003.

Bibliography

[HL07] Y. Hilewitz and R.B. Lee. Performing Advanced Bit Manipulations Efficiently in General-Purpose Processors. In *Proceedings of the 18th IEEE Symposium on Computer Arithmetic (ARITH-18)*, pages 251–260, Montpellier, France, 2007.

[HM03] H. C. Hunter and J. H. Moreno. A new look at exploiting data parallelism in embedded systems. In *CASES '03: Proceedings of the 2003 international conference on Compilers, architecture and synthesis for embedded systems*, pages 159–169, New York, NY, USA, 2003. ACM Press.

[HNBM01] A. Hoffmann, A. Nohl, G. Braun, and H. Meyr. A survey on modeling issues using the machine description language LISA. In *Acoustics, Speech, and Signal Processing, 2001. Proceedings. (ICASSP '01). 2001 IEEE International Conference on*, volume 2, pages 1137–1140, 2001.

[Hof02] A. Hoffmann. *Architecture Exploration for Embedded Processors with LISA*. PhD thesis, RWTH Aachen, 2002.

[HtB03] B. M. Hochwald and S. ten Brink. Achieving near-capacity on a multiple-antenna channel. *IEEE Trans. Commun.*, 51(3):389–399, March 2003.

[IEE09a] IEEE. IEEE Standard for Information technology–Telecommunications and information exchange between systems–Local and metropolitan area networks–Specific requirements Part 11: Wireless LAN Medium Access Control (MAC) and Physical Layer (PHY) Specifications Amendment 5: Enhancements for Higher Throughput. *IEEE Std 802.11n-2009 (Amendment to IEEE Std 802.11-2007 as amended by IEEE Std 802.11k-2008, IEEE Std 802.11r-2008, IEEE Std 802.11y-2008, and IEEE Std 802.11w-2009)*, pages c1 –502, 29 2009.

[IEE09b] IEEE Computer Society and IEEE Microwave Theory and Techniques Society. *IEEE Standard for Local and Metropolitan Area Networks Part 16: Air Interface for Broadband Wireless Access Systems*, IEEE Std 802.16-2009 edition, May 2009.

[JBL$^+$08] M. Joham, L.G. Barbero, T. Lang, W. Utschick, J. Thompson, and T. Ratnarajah. FPGA implementation of MMSE metric based efficient near-ML detection. In *Smart Antennas, 2008. WSA 2008. International ITG Workshop on*, pages 139 –146, 26-27 2008.

[JBOT09] J. Jaldén, L.G. Barbero, B. Ottersten, and J.S. Thompson. The Error Probability of the Fixed-Complexity Sphere Decoder. *Signal Processing, IEEE Transactions on*, 57(7):2711–2720, July 2009.

[JGKM04] S. Jintukar, J. Glossner, V. Kotlyar, and M. Moudgill. The Sandblaster automatic multithreaded vectorizing compiler. In *2004 Global Signal Processing Expo (GSPx) and International Signal Processing Conference (ISPC)*, Santa Clara, California, 2004.

Bibliography

[JKMR07] C.-P. Jeannerod, H. Knochel, C. Monat, and G. Revy. Faster floating-point square root for integer processors. In *Industrial Embedded Systems, 2007. SIES '07. International Symposium on*, pages 324 –327, 4-6 2007.

[JSJ09] J. Janhunen, O. Silvén, and M. Juntti. Comparison of the Software Defined Radio Implementations of the K-best List Sphere Detection. In *17th European Signal Processing Conference (EUSIPCO 2009)*, pages 2396–2400, Glasgow, Scotland, August 24-28 2009. EURASIP.

[KF98] F.R. Kschischang and B.J. Frey. Iterative decoding of compound codes by probability propagation in graphical models. *Selected Areas in Communications, IEEE Journal on*, 16(2):219 –230, Feb. 1998.

[KFA+07] M. Keating, D. Flynn, R. Aitken, A. Gibbons, and K. Shih. *Low Power Methodology Manual: For System-on-Chip Design (Integrated Circuits and Systems)*. Springer, 2007.

[Kno05] S. Knowles. The SoC Future is Soft. IEE Cambridge Branch Seminar 2005 http://devel.iee-cambridge.org.uk/arc/seminar05/slides/SimonKnowles.pdf, Dec. 2005.

[KSAF07] R. Klemm, J.P. Sabugo, H. Ahlendorf, and G. Fettweis. Using LISATek for the Design of an ASIP Core including Floating-Point Operations. In *Proceedings of the 10th GI/ITG/GMM Workshop "Methoden und Beschreibungssprachen zur Modellierung und Verifikation von Schaltungen und Systemen" (MBMV'07)*, Mar. 2007.

[KW08] T.-C. Kuo and A.N. Willson. A flexible decoder IC for WiMAX QC-LDPC codes. In *Custom Integrated Circuits Conference, 2008. CICC 2008. IEEE*, pages 527 –530, 21-24 2008.

[Lö04] O. Lüthje. *A Methodology for Automated Anaylsis of Application Specific Processor Models with Respect to Test Generation*. PhD thesis, RWTH Aachen, 2004.

[LBL+08] M. Li, B. Bougard, E.E. Lopez, A. Bourdoux, D. Novo, L. Van Der Perre, and F. Catthoor. Selective Spanning with Fast Enumeration: A Near Maximum-Likelihood MIMO Detector Designed for Parallel Programmable Baseband Architectures. In *Communications, 2008. ICC '08. IEEE International Conference on*, pages 737–741, May 2008.

[LCB+06] A. Lodi, A. Cappelli, M. Bocchi, C. Mucci, M. Innocenti, C. De Bartolomeis, L. Ciccarelli, R. Giansante, A. Deledda, F. Campi, M. Toma, and R. Guerrieri. XiSystem: a XiRisc-based SoC with reconfigurable IO module. *Solid-State Circuits, IEEE Journal of*, 41(1):85 – 96, Jan. 2006.

[LCM09] H. Lee, C. Chakrabarti, and T. Mudge. A Low-Power DSP for Wireless Communications. *Very Large Scale Integration (VLSI) Systems, IEEE Transactions on*, PP(99):13, 2009.

[Lec09]   M. Lechtenberg. Modellierung eines SIMD-DSPs mit konfigurierbarer Vektorlänge. Diploma thesis (German), Technische Universität Dortmund, Circuits and Systems Lab, Oct. 2009.

[LFN+09]   M. Li, R. Fasthuber, D. Novo, B. Bougard, L. Van Der Perre, and F. Catthoor. Algorithm-architecture co-design of soft-output ML MIMO detector for parallel application specific instruction set processors. In *Design, Automation Test in Europe Conference Exhibition, 2009. DATE '09.*, pages 1608–1613, 20-24 2009.

[Lin08]   Y. Lin. *Realizing Software Defined Radio – A Study in Designing Mobile Supercomputers*. PhD thesis, University of Michigan, 2008.

[LLW+06]   Y. Lin, H. Lee, M. Woh, Y. Harel, S. Mahlke, T. Mudge, C. Chakrabarti, and K. Flautner. SODA: A Low-power Architecture For Software Radio. In *Proc. 33rd Intl. Symposium on Computer Architecture (ISCA)*, 2006.

[LLW+07]   Y. Lin, H. Lee, M. Woh, Y. Harel, S. Mahlke, T. Mudge, C. Chakrabarti, and K. Flautner. SODA: A High-performance DSP Architecture for Software-Defined Radio. *IEEE Micro*, 27(1):114–123, Jan/Feb 2007.

[LTC03]   A. Lodi, M. Toma, and F. Campi. A pipelined configurable gate array for embedded processors. In *FPGA '03: Proceedings of the 2003 ACM/SIGDA eleventh international symposium on Field programmable gate arrays*, pages 21–30, New York, NY, USA, 2003. ACM.

[Mac99]   D.J.C. MacKay. Good error-correcting codes based on very sparse matrices. *Information Theory, IEEE Transactions on*, 45(2):399–431, Mar. 1999.

[MB06]   T. Mohsenin and B.M. Baas. Split-Row: A Reduced Complexity, High Throughput LDPC Decoder Architecture. In *Computer Design, 2006. ICCD 2006. International Conference on*, pages 320–325, Oct. 2006.

[MG08]   M. Moudgill and J. Glossner. The Sandblaster 2.0 Architecture and SB3500 Implementation. In *Proceedings of the Software Defined Radio Technical Forum (SDR Forum '08)*, Oct. 2008.

[MLG06]   H. G. Myung, J. Lim, and D. J. Goodman. Single carrier FDMA for uplink wireless transmission. *Vehicular Technology Magazine, IEEE*, 1(3):30–38, 2006.

[MMF09]   B. Mennenga, E. Matus, and G. Fettweis. Vectorization of the sphere detection algorithm. In *Circuits and Systems, 2009. ISCAS 2009. IEEE International Symposium on*, pages 2806–2809, May 2009.

[MN95]   D. J. C. MacKay and R. M. Neal. Good Codes based on Very Sparse Matrices. In *Cryptography and Coding, 5th IMA Conference*, number 1025 in Lecture Notes on Computer Science, pages 100–111, 1995.

Bibliography

[MN97]　　D.J.C. MacKay and R.M. Neal. Near Shannon limit performance of low density parity check codes. *Electronics Letters*, 33(6):457 –458, 13 1997.

[MS03]　　M.M. Mansour and N.R. Shanbhag. High-throughput LDPC decoders. *Very Large Scale Integration (VLSI) Systems, IEEE Transactions on*, 11(6):976–996, Dec. 2003.

[MVV+03]　B. Mei, S. Vernalde, D. Verkest, H. De Man, and R. Lauwereins. ADRES: An Architecture with Tightly Coupled VLIW Processor and Coarse-Grained Reconfigurable Matrix. In *Field-Programmable Logic and Applications 13th International Conference, FPL 2003*, 2003.

[MZBF09]　D.L. Milliner, E. Zimmermann, J.R. Barry, and G. Fettweis. A Fixed-Complexity Smart Candidate Adding Algorithm for Soft-Output MIMO Detection. *IEEE Journal of Selected Topics in Signal Processing*, 3(6):1016 –1025, Dec. 2009.

[MZS+03]　J. H. Moreno, V. Zyuban, U. Shvadron, F. D. Neeser, J. H. Derby, M. S. Ware, K. Kailas, A. Zaks, A. Geva, S. Ben-David, S. W. Asaad, T. W. Fox, D. Littrell, M. Biberstein, D. Naishlos, and H. Hunter. An innovative low-power high-performance programmable signal processor for digital communications. *IBM Journal of Research and Development*, 47(2.3):299 –326, Mar. 2003.

[Neu04]　　Y. Neuvo. Cellular phones as embedded systems. *IEEE International Solid-State Circuits Conference, 2004. Digest of Technical Papers. ISSCC. 2004*, 1:32 – 37, Feb. 2004.

[Nil07]　　A. Nilsson. *Design of programmable multi-standard baseband processors*. PhD thesis, Linköping University, Department of Electrical Engineering, 2007.

[NRZ06]　　D. Nuzman, I. Rosen, and A. Zaks. Auto-vectorization of interleaved data for SIMD. In *PLDI '06: Proceedings of the 2006 ACM SIGPLAN conference on Programming language design and implementation*, pages 132—143, New York, NY, USA, 2006. ACM Press.

[NTL09]　　A. Nilsson, E. Tell, and D. Liu. An $11\,mm^2$, 70 mW Fully Programmable Baseband Processor for Mobile WiMAX and DVB-T/H in $0.12\mu m$ CMOS. *Solid-State Circuits, IEEE Journal of*, 44(1):90 –97, Jan. 2009.

[NZ06]　　D. Nuzman and A. Zaks. Autovectorization in GCC — two years later. In *Proceedings of the GCC Developers' Summit*, June 2006.

[Ond05]　　S. Onder. An introduction to Flexible Architecture Simulation Tool (FAST) and Architecture Description Language ADL. Technical Report Technical Report TR 05-01, Michigan Technological University, 2005.

[Par80]　　D. S. Parker. Notes on Shuffle/Exchange-Type Switching Networks. *IEEE Trans. Comput.*, 29(3):213–222, 1980.

Bibliography

[Pea77]   M. C. Pease. The Indirect Binary n-Cube Microprocessor Array. *IEEE Trans. Comput.*, 26(5):458–473, 1977.

[Pee02]   S. L. A. Pees. *Modeling Embedded Processors and Generating Fast Simulators Using the Machine Description Language LISA*. PhD thesis, RWTH Aachen, 2002.

[PHZM99]   S. Pees, A. Hoffmann, V. Zivojnovic, and H. Meyr. LISA–machine description language for cycle-accurate models of programmable DSP architectures. In *DAC '99: Proceedings of the 36th ACM/IEEE conference on Design automation*, pages 933–938, New York, NY, USA, 1999. ACM.

[Pit97]   N. P. Pitsianis. A Kronecker Compiler for Fast Transform Algorithms. In *Proceedings of the Eighth SIAM Conference on Parallel Processing for Scientific Computing, PPSC 1997*, January 1997.

[PK98]   M. K. Prasad and R. K. Kolagotla. Modulo address generators for DSPs. *IEE Electronics Letters*, 34(17):1653–1654, Aug. 1998.

[PLGG01]   J. Van Praet, D. Lanneer, W. Geurts, and G. Goossens. Processor Modeling and Code Selection for Retargetable Compilation. *ACM Transactions on Design Automation of Electronic Systems*, 6(3):1–30, 2001.

[PMJ+05]   M. Püschel, J.M.F. Moura, J.R. Johnson, D. Padua, M.M. Veloso, B.W. Singer, J. Xiong, F. Franchetti, A. Gacic, Y. Voronenko, K. Chen, R.W. Johnson, and N. Rizzolo. SPIRAL: Code Generation for DSP Transforms. *Proceedings of the IEEE*, 93(2):232 –275, Feb. 2005.

[Poh81]   M. Pohst. On the computation of lattice vectors of minimal length, successive minima and reduced bases with applications. *SIGSAM Bull.*, 15(1):37–44, 1981.

[PT09]   T. Pitkanen and J. Takala. Low-power application-specific processor for FFT computations. In *Acoustics, Speech and Signal Processing, 2009. ICASSP 2009. IEEE International Conference on*, pages 593 –596, 19-24 2009.

[Pul08]   D. Pulley. Multi-core DSP for base stations: Large and small. In *Design Automation Conference, 2008. ASPDAC 2008. Asia and South Pacific*, pages 389 –391, 21-24 2008.

[Rad68]   C. M. Rader. Discrete Fourier transforms when the number of data samples is prime. *Proc. IEEE*, 56(6):1107–1108, June 1968.

[Rai06]   S. K. Raina. *FLIP: A Floating-Point Library for Integer Processors*. PhD thesis, l'École Normale Superieure de Lyon, 2006.

[Ram07]   U. Ramacher. Software-defined radio prospects for multistandard mobile phones. *Computer*, 40(10):62–69, 2007.

[RdBKC06] P. Radosavljevic, A. de Baynast, M. Karkooti, and J. R. Cavallaro. High-Throughput Multi-rate LDPC Decoder based on Architecture-Oriented Parity Check Matrices. In *European Signal Processing Conference 2006 (EUSIPCO 2006)*, 2006.

[RDK+00] S. Rixner, W.J. Dally, B. Khailany, P. Mattson, U.J. Kapasi, and J.D. Owens. Register organization for media processing. In *High-Performance Computer Architecture, 2000. HPCA-6. Proceedings. Sixth International Symposium on*, pages 375–386, 2000.

[Rep08] Rep. ITU-R M.2134. Requirements related to technical performance for IMT-Advanced radio interface(s). Technical report, International Telecommunication Union - Radiocommunication, 2008.

[RMR+07] P. Raghavan, S. Munaga, E. Rey Ramos, A. Lambrechts, M. Jayapala, F. Catthoor, and D. Verkest. A Customized Cross-Bar for Data-Shuffling in Domain-Specific SIMD Processors. In *Proc. of Architecture and Computing Systems (ARCS)*, 2007.

[RS03] M. Ros and P. Sutton. Compiler Optimization and Ordering Effects on VLIW Code Compression. In *International Conference on Compilers, Architecture and Synthesis for Embedded Systems*, 2003.

[RSU01] T.J. Richardson, M.A. Shokrollahi, and R.L. Urbanke. Design of capacity-approaching irregular low-density parity-check codes. *Information Theory, IEEE Transactions on*, 47(2):619 –637, Feb. 2001.

[RU01] T.J. Richardson and R.L. Urbanke. Efficient encoding of low-density parity-check codes. *Information Theory, IEEE Transactions on*, 47(2):638 –656, Feb. 2001.

[Rus78] R. M. Russell. The CRAY-1 computer system. *Commun. ACM*, 21(1):63–72, 1978.

[San09] Sandbridge Technologies. Sandbridge announces certified BDTI Communications Benchmark (OFDM) results for its Sandblaster SB3500. Press release http://www.sandbridgetech.com/pdf/sb_PR_BDTI_v1_8_FINAL.pdf, Jan. 2009.

[SBM+04] G. L. Stüber, J. R. Barry, S. W. McLaughlin, L. Ye, M. A. Ingram, and T. G. Pratt. Broadband MIMO-OFDM wireless communications. *Proc. IEEE*, 92(2):271– 294, Feburary 2004.

[SC08] Y. Sun and J.R. Cavallaro. A low-power 1-Gbps reconfigurable LDPC decoder design for multiple 4G wireless standards. In *SOC Conference, 2008 IEEE International*, pages 367 –370, 17-20 2008.

[SE94] C. P. Schnorr and M. Euchner. Lattice basis reduction: Improved practical algorithms and solving subset sum problems. *Mathematical Programming*, 66(1):181 – 199, 1994.

[SE09] ST-Ericsson. Low-power embedded vector DSP — EVP VD32041 32-bit embedded-vector processor for SoCs. Online http://www.stericsson.com/sales_marketing_resources/VD32041BR_1.pdf, Feb. 2009.

Bibliography

[Sie79]   H. J. Siegel. Interconnection Networks for SIMD Machines. *Computer, Special Issue on Circuit Switching*, 12(6):57–69, 1979.

[Sin67]   R. C. Singleton. On computing the fast Fourier transform. *Commun. ACM*, 10(10):647–654, 1967.

[SLG04]   E. Sharon, S. Litsyn, and J. Goldberger. An efficient message-passing schedule for LDPC decoding. In *Electrical and Electronics Engineers in Israel, 2004. Proceedings. 2004 23rd IEEE Convention of*, pages 223–226, Sept. 2004.

[SLG07]   E. Sharon, S. Litsyn, and J. Goldberger. Efficient Serial Message-Passing Schedules for LDPC Decoding. *Information Theory, IEEE Transactions on*, 53(11):4076–4091, Nov. 2007.

[SM06]   L. Schwoerer and K. Moerman. Benchmarking MIMO OFDM algorithms on the EVP. In *Proceedings of the GSPx 2006 Pervasive Signal Processing Conference*, Santa Clara, CA, USA, Oct.-Nov. 2006.

[SMN+09]   V. Surducan, M. Moudgill, G. Nacer, E. Surducan, P. Balzola, J. Glossner, S. Stanley, Meng Yu, and D. Iancu. The Sandblaster Software-Defined Radio Platform for Mobile 4G Wireless Communications. *International Journal of Digital Multimedia Broadcasting*, 2009:9, 2009.

[Smo02]   M. Smotherman. Understanding EPIC Architectures and Implementations. In *40th Annual ACM Southeast Conference, Raleigh*, pages 71–78, April 2002.

[SMPTM79]   H. J. Siegel, R. J. McMillen, and Jr. P. T. Mueller. A Survey of Interconnection Methods for Reconfigurable Parallel Processing Systems. In *AFIPS Conference Proceedings Volume 48: 1979 National Computer Conference*, 1979.

[SMZC07]   S. Seo, T. Mudge, Y. Zhu, and C. Chakrabarti. Design and analysis of LDPC decoders for software defined radio. In *Signal Processing Systems, 2007 IEEE Workshop on*, pages 210–215, Oct. 2007.

[SPVB08]   R. Seindal, F. Pinard, G. V. Vaughan, and E. Blake. *GNU M4, version 1.4.12 - A powerful macro processor*, 1.4.12 edition, Sept. 2008.

[SS78]   H. J. Siegel and S. D. Smith. Study of multistage SIMD interconnection networks. In *ISCA '78: Proceedings of the 5th annual symposium on Computer architecture*, pages 223–229, New York, NY, USA, 1978. ACM.

[SVPG+10]   A. Squires, B. Van Poucke, J. Glossner, K. Moerman, K. van Berkel, and M. Uhm. Commercial Baseband Technology Overview: The Current State of Technology Development and Future Directions. Technical Report WINNF-09-P-0009-V1.0.0, Wireless Innovation Forum, Feb. 2010.

Bibliography

[SYM08] X. Shi, S. Yoshizawa, and Y. Miyanaga. Performance evaluation of quasi-cyclic LDPC codes for IEEE802.11n based MIMO-OFDM systems. In *Communications, Control and Signal Processing, 2008. ISCCSP 2008. 3rd International Symposium on*, pages 1330 –1333, 12-14 2008.

[Syn07] Synopsys. Synopsys Low-Power Solution. White Paper, available at http://electronics.wesrch.com/pdf_file/SE1_1191538495/low-power-design-solution.pdf, June 2007.

[Syn09a] Synopsys. *Design Compiler User Guide*, C-2009.06 edition, June 2009.

[Syn09b] Synopsys. *Power Compiler User Guide*, C-2009.06-SP2 edition, Sept. 2009.

[SZLW07] X.-Y. Shih, C.-Z. Zhan, C.-H. Lin, and A.-Y. Wu. A 19-mode $8.29\,mm^2$ 52-mW LDPC Decoder Chip for IEEE 802.16e System. In *VLSI Circuits, 2007 IEEE Symposium on*, pages 16 –17, 14-16 2007.

[Tai06] Taiwan Semiconductor Manufacturing Company, Ltd. *TCBN90GTHP TSMC 90nm Core Library Databook*, 1.1 edition, Dec. 2006.

[Tan81] R. Tanner. A recursive approach to low complexity codes. *Information Theory, IEEE Transactions on*, 27(5):533 – 547, Sept. 1981.

[Tec06] Technical Specification Group Radio Access Network. TR 25.913 Requirements for Evolved UTRA (E-UTRA) and Evolved UTRAN (E-UTRAN) (Release 7) . Technical Report V7.3.0, 3rd Generation Partnership Project, Mar. 2006.

[Tec07] Technical Specification Group Radio Access Network. TS 25.213 Spreading and modulation (FDD) (Release 7). Technical Report Release 7, 3rd Generation Partnership Project, Sept. 2007.

[Tec09a] Technical Specification Group Radio Access Network. TR 36.913 Requirements for further advancements for Evolved Universal Terrestrial Radio Access (E-UTRA) (LTE-Advanced) (Release 9). Technical Report V9.0.0, 3rd Generation Partnership Project, Dec. 2009.

[Tec09b] Technical Specification Group Radio Access Network. TS 36.211 Evolved Universal Terrestrial Radio Access (E-UTRA); Physical Channels and Modulation (Release 8). Technical Report V8.9.0, 3rd Generation Partnership Project, Dec. 2009.

[Tec10] Technical Specification Group Radio Access Network. TS 36.101 Evolved Universal Terrestrial Radio Access (E-UTRA); User Equipment (UE) radio transmission and reception (Release 8). Technical Report V8.9.0, 3rd Generation Partnership Project, Mar. 2010.

[Tel99] I. E. Telatar. Capacity of multi-antenna Gaussian channels. *European Transactions on Telecommunications*, 10(6):585–596, 1999.

[Tem83] C. Temperton. Self-sorting mixed-radix fast Fourier transforms. *Journal of Computational Physics*, 52(1):1–23, Oct 1983.

[Ten05] Tensilica. Xtensa Architecture and Performance. Tensilica White Paper, online `http://www.tensilica.com/products/literature-docs/white-papers.htm`, Oct. 2005.

[Ten08] Tensilica. The What, Why, and How of Configurable Processors. Tensilica White Paper, online `http://www.tensilica.com/products/literature-docs/white-papers.htm`, Oct. 2008.

[TMAJ08] S. Thoziyoor, N. Muralimanohar, J. H. Ahn, and N. P. Jouppi. CACTI 5.1. Technical Report HPL-2008-20, HP Laboratories, Palo Alto, Palo Alto, April 2008.

[TSFB07] A. Tomasoni, M. Siti, M. Ferrari, and S. Bellini. Turbo-LORD: A MAP-Approaching Soft-Input Soft-Output Detector for Iterative MIMO Receivers. In *Global Telecommunications Conference, 2007. GLOBECOM '07. IEEE*, pages 3504 –3508, 26-30 2007.

[TSG06] TSG-RAN WG1 (Ericsson). DFT size for uplink transmissions. 3GPP Technical Document R1-062852, Oct. 2006.

[TSS+04] R.M. Tanner, D. Sridhara, A. Sridharan, T.E. Fuja, and Jr. Costello, D.J. LDPC block and convolutional codes based on circulant matrices. *Information Theory, IEEE Transactions on*, 50(12):2966 – 2984, Dec. 2004.

[VB99] E. Viterbo and J. Boutros. A universal lattice code decoder for fading channels. *IEEE Trans. Inform. Theory*, 45(5):1639–1642, July 1999.

[vHM+04] C. H. van Berkel, F. Heinle, P. P. E. Meuwissen, K. Moerman, and M. Weiss. Vector processing as an enabler for software-defined radio in handsets from 3G+WLAN onwards. In *Proceedings of the 2004 software-defined radio technical conference (SDR'04)*, Scottsdale, Arizona, U.S.A., Sept. 2004.

[vHM+05] K. van Berkel, F. Heinle, P. P. E. Meuwissen, K. Moerman, and M. Weiss. Vector processing as an enabler for software-defined radio in handheld devices. *EURASIP Journal on Applied Signal Processing*, 16:2613–2625, 2005.

[Wak68] A. Waksman. A permutation network. *J. ACM*, 15(1):159—163, 1968.

[WBAHS08a] P. Westermann, G. Beier, H. Ait-Harma, and L. Schwoerer. Developing FFTs for SC-FDMA on the Embedded Vector Processor. In *Proceedings of the 13th International OFDM-Workshop (InOWo'08)*, 2008.

[WBAHS08b] P. Westermann, G. Beier, H. Ait-Harma, and L. Schwoerer. Performance Analysis of W-CDMA Algorithms on a Vector DSP. In *Proceddings of the 4–th European Conference on Circuits and Systems for Communications (ECCSC'08)*, July 10–11 2008.

Bibliography

[WBAHS09] P. Westermann, G. Beier, H. Ait-Harma, and L. Schwoerer. Performance Analysis of Wireless Communication Algorithms on a Vector Signal Processor. *Rev. Roum. Sci. Techn.– Éectrotechn. et Énerg.*, 54(3):291–300, 2009.

[WC92] A. Wolfe and A. Chanin. Executing compressed programs on an embedded RISC architecture. In *MICRO 25: Proceedings of the 25th annual international symposium on Microarchitecture*, pages 81–91, Los Alamitos, CA, USA, 1992. IEEE Computer Society Press.

[WEL09] D. Wu, J. Eilert, and D. Liu. Evaluation of MIMO Symbol Detectors for 3GPP LTE Terminals. In *17th European Signal Processing Conference (EUSIPCO 2009)*, pages 2431–2435, Glasgow, Scotland, August 24-28 2009. EURASIP.

[WFGV98] P.W. Wolniansky, G.J. Foschini, G.D. Golden, and R.A. Valenzuela. V-BLAST: an architecture for realizing very high data rates over the rich-scattering wireless channel. In *Signals, Systems, and Electronics, 1998. ISSSE 98. 1998 URSI International Symposium on*, pages 295 –300, 29 1998.

[Wib96] N. Wiberg. *Codes and Decoding on General Graphs*. PhD thesis, Linköping University, Sweden, 1996.

[WLS+08a] M. Woh, Y. Lin, S. Seo, S. Mahlke, T. Mudge, C. Chakrabarti, R. Bruce, Kershaw, A. D. Reid, M. Wilder, and K. Flautner. From SODA to Scotch: The Evolution of a Wireless Baseband Processor. In *Proc. 41st Intl. Symposium on Microarchitecture (MICRO)*, pages 152–163, Nov. 2008.

[WLS+08b] M. Woh, Y. Lin, S. Seo, T. Mudge, and S. Mahlke. Analyzing the scalability of SIMD for the next generation software defined radio. In *IEEE International Conference on Acoustics, Speech and Signal Processing 2008, ICASSP 2008.*, pages 5388–5391, March 31 2008–April 4 2008 2008.

[WMMC10] M. Woh, S. Mahlke, T. Mudge, and C. Chakrabarti. Mobile Supercomputers for the Next-Generation Cell Phone. *Computer*, 43(1):81 –85, Jan. 2010.

[WS09a] P. Westermann and H. Schröder. Constraints on the SIMD Vectorization of Radix-2 and Mixed-Radix FFTs. In *17th European Signal Processing Conference (EUSIPCO 2009)*, pages 1274–1278, Glasgow, Scotland, August 24-28 2009. EURASIP.

[WS09b] P. Westermann and H. Schröder. Modeling Scalable SIMD DSPs in LISA. In K. Bertels et al., editor, *SAMOS 2009, LNCS 5657*, pages 160–169. Springer-Verlag Berlin Heidelberg, 2009.

[WSK07] P. Westermann, L. Schwoerer, and A. Kaufmann. Applying Data Mapping Techniques to Vector DSPs. In *Proceedings 2007 International Conference on Embedded Computer Systems: Architectures, Modeling and Simulation (IC-SAMOS 2007)*, pages 1–8, Samos, Greece, July 2007.

[WSK08]   P. Westermann, L. Schwoerer, and A. Kaufmann. Applying Data Mapping Techniques to Vector DSPs. *Journal of VLSI Signal Processing Systems*, Special Issue on SAMOS'2007:57–72, 2008.

[WSL+07]   M. Woh, S. Seo, H. Lee, Y. Lin, S. Mahlke, T. Mudge, C. Chakrabarti, and K. Flautner. The next generation challenge for software defined radio. In *Proc. 7th Intl. Workshop on Systems, Architectures, Modeling, and Simulation (SAMOS)*, 2007.

[WSM+09]   M. Woh, S. Seo, S. Mahlke, T. Mudge, C. Chakrabarti, and K. Flautner. AnySP: anytime anywhere anyway signal processing. In *Proc. 36th Intl. Symposium on Computer Architecture (ISCA)*, volume 37, pages 128–139, New York, NY, USA, 2009. ACM.

[WSM+10]   M. Woh, S. Seo, S. Mahlke, T. Mudge, C. Chakrabarti, and K. Flautner. AnySP: Anytime Anywhere Anyway Signal Processing. *Micro, IEEE*, 30(1):81 –91, Jan.-Feb. 2010.

[WTCM02]   K.-W. Wong, C.-Y. Tsui, R.S.-K. Cheng, and W.-H. Mow. A VLSI architecture of a K-best lattice decoding algorithm for MIMO channels. In *Circuits and Systems, 2002. ISCAS 2002. IEEE International Symposium on*, volume 3, pages III–273 – III–276 vol.3, 2002.

[WTW09]   X. Wu, J. S. Thompson, and A. M. Wallace. An Improved Sphere Decoding Scheme for MIMO Systems using an Adaptive Statistical Threshold. In *17th European Signal Processing Conference (EUSIPCO 2009)*, pages 2668–2672, Glasgow, Scotland, August 24-28 2009. EURASIP.

[XWL02]   Y. Xie, W. Wolf, and H. Lekatsas. Code Compression for VLIW Processors Using Variable-to-fixed Coding. In *Proceedings of Fifteenth International Symposium on System Synthesis (ISSS 2002)*, 2002.

[XWL06]   Y. Xie, W. Wolf, and H. Lekatsas. Code Compression for Embedded VLIW Processors Using Variable-to-Fixed Coding. *IEEE Trans. VLSI Syst.*, 14(5):525–536, May 2006.

[ZC09]   Y. Zhu and C. Chakrabarti. Architecture-aware LDPC code design for multiprocessor software defined radio systems. *IEEE Trans. Signal Processing*, 57(9):3679–3692, 2009.

[ZPM96]   V. Zivojnovic, S. Pees, and H. Meyr. LISA-machine description language and generic machine model for HW/SW co-design. In *VLSI Signal Processing, IX, 1996., [Workshop on]*, pages 127 –136, Oct. /Nov. 1996.

I want morebooks!

Buy your books fast and straightforward online - at one of world's fastest growing online book stores! Environmentally sound due to Print-on-Demand technologies.

Buy your books online at
www.morebooks.shop

Kaufen Sie Ihre Bücher schnell und unkompliziert online – auf einer der am schnellsten wachsenden Buchhandelsplattformen weltweit! Dank Print-On-Demand umwelt- und ressourcenschonend produziert.

Bücher schneller online kaufen
www.morebooks.shop

KS OmniScriptum Publishing
Brivibas gatve 197
LV-1039 Riga, Latvia
Telefax +371 686 204 55

info@omniscriptum.com
www.omniscriptum.com

Printed by Books on Demand GmbH, Norderstedt / Germany